A History of MONTANA AGRICULTURE

A LIFE OF DISCOVERY

JODY L. LAMP & MELODY DOBSON

Published by The History Press
Charleston, SC
www.historypress.com

Copyright © 2021 by Jody L. Lamp and Melody Dobson
All rights reserved

Back cover, top: Three generations of Arthur Langmans, circa 1950, from left to right: Arthur Jerome "Jerry" (born August 24, 1918 in Grand Island, NE); young Arthur "Scott" Langman (born April 2, 1948 in Billings, MT); and Arthur Henry Langman (born March 20, 1882 in Grand Island, NE), co-founder of the Grand Island, Nebraska's Horse & Mule Market and founder of Billings Livestock Commission at 1202 First Avenue North, Billings, Montana. *Courtesy of the Scott Langman family.*
Back cover, bottom: *Morning Mail* by artist Gene Roncka, Willow Point Gallery, Ashland, Nebraska, reminds coauthor Melody Dobson of her own upbringing in northeast Montana of walking down the lane to the mailbox. "Mail was a special event, and in the rural community of Vida it was delivered only three days a week," said Dobson. "It's that essence of yesteryear with families living and working together that symbolizes the generational impact Montanans have made to make agriculture the number one industry in the state."

First published 2021

Manufactured in the United States

ISBN 9781467136501

Library of Congress Control Number: 2021931097

Notice: The information in this book is true and complete to the best of our knowledge. It is offered without guarantee on the part of the authors or The History Press. The authors and The History Press disclaim all liability in connection with the use of this book.

All rights reserved. No part of this book may be reproduced or transmitted in any form whatsoever without prior written permission from the publisher except in the case of brief quotations embodied in critical articles and reviews.

Meadowlark, by Gene Roncka

Front cover: *Meadowlark* depicts a sunny day in an era when our homesteading grandmothers' daily routine may have included draping and pegging the washed laundry to dry on a clothesline raised between two poles strong enough to withstand the prairie winds. To grandmother's listening enjoyment, the western meadowlark serenades and repeatedly echoes to her a "whistle while I work" melody, a familiar tune recognized throughout the High Plains.

The meadowlark was chosen as the state bird of Montana in 1930 by a vote of the schoolchildren of the state. By an act of the Montana legislature, the western meadowlark was adopted as the official state bird on March 14, 1931. The act reads, in part:

Section 1. The bird known as the Western Meadow Lark, Sturnella-Neglecta (Audubon) as preferred by a referendum vote of Montana school children, shall be designated and declared to be the official bird of the State of Montana.[1]

As agriculture ambassadors and advocates, we dedicate this book to preserving the historical narrative of our homesteading families before generations removed from the family farm forget their ancestral roots.

Contents

Foreword, by Barb Skelton — 9
Acknowledgements — 11
Prologue — 15

Part 1. Creating Montana's Agriculture: Spaces and Places
1. The Treasure State's Agriculture Foundation — 19
2. The Hunt for Gold Fosters Agriculture Growth — 25
3. Pioneering Responsible Citizenship — 33
4. Montana's Early Agriculture Blooms and Booms — 36
5. They Came West to Build a New Future: The Kents — 41
6. At Home on the Judith: The Hardenbrooks and the Skeltons — 52
7. Homesteading in Northeastern Montana: Alfred and Regina Wanderaas — 60
8. The Homesteading Boom — 76

Part 2. Growing Montana's Agriculture: Inventions and Commodities
9. The Miracle Invention — 83
10. The Mechanical Revolution — 87
11. Learning to Grow in Diverse Conditions — 94
12. The Legacy Continues: Arthur H. Langman and the Billings Livestock Commission — 102
13. From Copper King to the Sport of Kings: Montana's Connection to the Thoroughbred Horse Racing Industry — 136
14. New Commodities Continue to Be Introduced and Succeed — 146

Contents

Part 3. Experiencing Montana's Agriculture: Events of Significance
15. The Cooperative Model   159
16. A Personal Perspective: The Jake Frank Story   168
17. Weather   182

Conclusion   191
Notes   195
About the Authors   207

FOREWORD

It is a privilege to introduce *A History of Montana Agriculture: A Life of Discovery*, a book that Melody Dobson and Jody Lamp have worked hard to publish, offering historical highlights of Montana's number one industry: agriculture. This dynamic duo has worked on "opening the door" to agriculture's importance since 2015, when they established the American Doorstop Project, hoping to portray how agriculture is a great gift to our nation and the world. In this edition, they unearth the riches of the Treasure State, my home state and the birthplace of my family for six generations.

The importance of preserving our agricultural heritage—the foundation of our nation—has never been more needed. Fewer of us in agriculture feed more people than ever before. The pressure to maintain our stewardship of the land and increase food production has never been more challenging. But, as a fifth-generation Montanan, I appreciate that challenge as part and parcel of my heritage. My daughter Abbie, who represents a sixth generation of my family, whose ranching is anchored in the heart of Charlie Russell Country, cradled near the Little Belt Mountains, is mindful of our legacy in this Big Sky land. Her ancestors, my great-great-grandparents, homesteaded the Hardenbrook Ranch in 1887 upon arriving in Great Falls. The Skelton family spent time with the legendary artist Charlie Russell, whose work captured the rugged life of those early ranchers and homesteaders. Indeed, Russell's painting *The First Furrow* depicts Abbie's great-grandfather William Skelton. That painting is no longer ours, since it was used to pay the doctor's bill when the children, stricken by chickenpox, needed medical services.

Foreword

There are a lot of "great-great-greats" in our family, and they are all great since they buckled down to create a life that not only sustained them but also helped build a life for future family. But sometimes one wonders what was great about kneeling in mud and dirty straw in one's pajamas to get a newborn to suck. Was it great to pull a lamb and rush to church with slobber and slime still on your hands? The pastor didn't mind. "Just glad you could come," he said. Was it great to shed manure-caked dungarees and hurry to make it to the CMR Art Auction, where dealers gathered from all over the world, and discover you still had a clot of dirt on the hem of your fancy dress (sewn on your Singer)?

And many of us worked without a permanent stake in the ranch or farm. Our names were absent from the deed. Once I considered selling my portion—because my name was on the deed—to save my daughter all this toil. But the answer came easy. No amount of cash could replace the sunsets we've seen with sandhill cranes swooping by, Angus calves running and bucking in a green field or longhorns swinging their horns to protect their babies. Cash can't buy riding home to a steak on the grill and baked spuds from the garden. Cash won't make up for a good neighbor stopping for coffee or a whiskey at the end of a good, God-given day.

I am particularly proud to be a woman in agriculture representing the long lineage of women who worked as hard and often harder than their male counterparts. Not only did we night calf, but we also cooked the meals for the hands. Not only did we nurse bum lambs, but we also nursed our family. We cleaned house and shoveled out the chicken house. We roped, branded and herded cattle. We saddle-broke horses and hauled hay with our tractors. We are still part mechanic, part seamstress, bookkeeper and beekeeper. Whatever is needed, we do. Our only manicure comes from the grit grinding down our nails.

My daughter knows, as I did and my grandmothers before me, the joys and the perils of the business we've chosen. We're Montana Proud. This book helps explain why.

ACKNOWLEDGEMENTS

Writing this book has been a tremendous experience, and without the love, support and encouragement of our families and friends, it would have never been accomplished. Each one of you has a special place in our hearts. We owe a debt of gratitude for everyone who graciously shared their stories and spent time looking through their scrapbooks and albums to find pictures, newspaper clippings and articles. We treasure those times and trust that the walk down memory lane was as pleasant for you as it was for us.

The community hospitality we experienced while being away from our homes and families made us feel welcomed. We love everything about rural America, including Montana's back roads, historical markers, accommodating rest areas with interpretive signage about the area and beautiful scenery.

Thank you to all the librarians who helped us find books, suggested titles, photographed copies and forgave our overdue fines. We implore you to keep sharing your love and knowledge of your local history.

As we enjoyed the many interpretive centers and museums, we want to especially acknowledge their boards of directors, executive directors, staffs and curators as they pulled articles and photos for us. The state and county historical societies provided invaluable insights, informative conversations and encouragement as they discovered our intentions to "write the book!"

To all of Montana's agritourism promoters, chambers of commerce, Rotarians, farm broadcasters, agriculture reporters and journalists,

Acknowledgements

agribusinesses and the industry of agriculture itself: your enthusiasm fuels our passion and makes us appreciate advocating for agriculture even more, as we realize that so many have not had the same experiences. As granddaughters and great-granddaughters of homesteaders, we consider our heritage a privilege and honor and do not take the memories of their hard work and love for granted.

The authors wish to acknowledge both Gene and Mary Roncka, Willow Point Gallery, Ashland, Nebraska, for sharing their time and treasures to adorn the front and back covers and pages of both our Nebraska and Montana agriculture history books. Montana also holds a special spot in Gene Roncka's heritage, as his homesteading roots are seeped deep in the sagebrush prairie of Charlie Russell country north of the Missouri River Breaks. As an artist, Roncka has dedicated a lifetime to exploring the landscapes of our nation and the stories behind the scenes. Choosing to capture life on the High Plains, the homesteading years are an important element of his work. Roncka was chosen to paint a mural for the Homestead National Historical Park's Education Center in Beatrice, Nebraska. Gene also created ornament designs of the pioneering years for the Homestead National Monument (now named Homestead National Historical Park) that were displayed on the White House Christmas tree in December 2007. Roncka's uncle, his mother's brother, acquired the homestead land south of Malta and it was held in the family for years until recently. From reminiscing and the stories passed down, Roncka attained a profound understanding of the homesteader's life and challenges in the open range land of central, Montana. Balanced with conversations and first-hand accounts from those who "proved-up" the first homesteads in Beatrice, Nebraska, and the nearby communities Roncka brings a perspective to the "days gone by" that emotionally moves the viewer down memory lane.

The American Doorstop Project co-founders and authors gratefully acknowledge the contributions of many individuals and organizations in making *A History of Montana Agriculture: A Life of Discovery* a reality. For their financial and in-kind support, we thank the following:

The Montana History Foundation
21st Century Equipment
Kelley Bean
Barb Skelton
SINGH Contracting Inc.
Stockman Bank

Acknowledgements

Dr. Steven Lutzwick
Montana State University Billings
Skip King; Billings, Laurel, Lockwood Ace Hardware
Felton Angus Ranch
Yellowstone Valley Electric Cooperative
Hill County Electric Cooperative
Triangle Communications
Nemont
George and Karen Yost
Tom and Lorie Pelatt
Lance Lanning
Drange Apiary
Alltech—Dawn Schooley
Zerbe Brothers
Mike Wilson—Whitewood Transport
Billings Livestock Commission
Scott Langman
Jann Parker and the late Bill Parker
Bob Cook and Bill Cook
The Michael Lamp Family
The Bernard and Ruth Wanderaas Family
The Philip and Iris Dobson Family
The John and Marilyn Vine Family
The Leland and Jean Hintz Family
Lee Ann and Richard Rush
Yellowstone Boys and Girls Ranch
Corcoran Trucking
Leonard and Kathy Dailey Jr.
George Mattson Farms—Carl Mattson
Huntley Project Museum of Irrigated Agriculture
Lynette Nealy Farmers Insurance
Rodney Broderson
Adult Resource Alliance of Yellowstone County
Photographic Solutions

Lastly, a special thanksgiving to our Heavenly Father for His hand of protection while we traveled and the provision of safety over our households while we were working on this project. From the bottom of our grateful hearts, we thank you!

PROLOGUE

JODY L. LAMP AND MELODY DOBSON

The big idea about Montana is that it is, well…BIG. In this narrative, *A History of Montana Agriculture: A Life of Discovery*, we work to capture the essence of that grandeur and all the state offers. Our focus and muse center on agriculture as a reminder of Montana's rich and diverse history. Agriculture embodies our roots, memories and antiquity. It's our belonging and connection to some of Montana's earliest inhabitants, who symbolized an agrarian existence.

As granddaughters and great-granddaughters of homesteaders, there is a sense of pride and dedication to remember, share and advocate for the state's number one industry—agriculture. Our families, friends and neighbors have been involved in and enthralled with this lifestyle for generations.

While we can't tell every family's story, we can share stories that every family can relate to. Each of us has a story, and we encourage you to capture it. Whether looking at old photographs, reading a letter written by your great-great-grandparent or perusing a newspaper clipping depicting the worst winter storm of the decade or century, take time to reminisce. Your generational and geological connections may bring detailed events to light for others to bond their own stories together. Whether your stories have become tall tales embedded in fact but embellished with colorful adjectives and exclamation points, memories are an important link to our space, place and time.

As co-founders and authors of the American Doorstop Project, our intent is to capture the heritage that built the doorframes of our past, allowing us to walk in the present and into our future. We offer a window into yesterday's activity to welcome an understanding of our early beginnings and the appreciation and discovery of what it has taken to get to where we are today.

Prologue

Back Roads brings back memories of days gone by and the reminder to get out and discover the back roads of Montana. *Courtesy of Gene Roncka, Willow Point Gallery, Ashland, Nebraska.*

Our history matters, because matters of history help give us definition and serve as building blocks for strengthening character and creating integrity.

It's only by respecting our past that we can formulate decisions on how to preserve our legacies and narratives. Many who have grown up in rural settings continue the tradition of the family farm and ranch, while others work in industry-related businesses or agriculture support agencies. Others may serve on business, foundation or museum boards of directors or serve in political office. A relationship of respect and integrity for our cultural heritage helps to serve as corner posts and compass points guiding us in understanding the past, present and future.

If you travel in Montana, you quickly understand just how BIG the Treasure State is and how overwhelming it can be. The word *divide* takes on a new meaning in Montana, and phrases like "just up the road" are open to interpretation and need definition. Trust us. Open-ended statements like "over on the other side of the hill" could mean more than a mile or two. A good sense of geography is essential in exploring the state. It helps to know the back roads and byways so that you reach your intended destination. Make a note, there isn't always a gas station waiting for you or a convenience store at that "dot on the map" that looked like a town. Sometimes that dot represents a significant period in Montana's agriculture history.

Part 1
CREATING MONTANA'S AGRICULTURE: SPACES AND PLACES

1
THE TREASURE STATE'S AGRICULTURE FOUNDATION

From the walking paths of Traveler's Rest State Park, near Lolo, Montana, one can pause and reflect on the years of discovery that settled the Bitterroot Valley, known as the state's birthplace and the founding of its agriculture. A National Historic Landmark of the Lewis and Clark Expedition, Travelers Rest serves as an intersection of history. The Corps of Discovery camped here to make preparations for the journey over the great Rocky Mountains and stopped again on their return trip. Their journals and maps were the first to provide insights and the location of the fertile area south of Missoula on U.S. Highway Route 93.

Native Americans from neighboring tribes traveled the Bitterroot Valley, home to the Salish Indians, on their way to hunt at the buffalo grounds on the plains. The Salish, also known as the Flathead Indians, raised horses in the rich, fertile valley that were "greatly admired and coveted by the other Indians."[2] They met the Lewis and Clark Expedition in September 1805, offered them food and traded horses to the explorers for their mountainous journey ahead. Captain Clark deemed the horses "elegant."[3]

For the Salish, protecting their horses from the continual raids by enemy tribes, especially the Blackfeet, was a constant challenge. Having learned of the "powerful medicine" of the "Black Robes" from the Iroquois Indians of Canada, the Salish wanted the medicine to help them in protecting their horses and way of life. While hunting and trapping on Salish lands, the Iroquois shared the stories taught to them by the black-robed Jesuit priests about "the Christian God, the Great Spirit and his powerful medicine."[4]

Along with desiring the power of the white man's God, they also heard of the food the "fathers" grew out of the land. It was becoming more difficult for them to get to the buffalo grounds because of the Blackfoot tribe's control. Their neighbors, the Nez Perce, also desired to have "black robes" of their own.[5] With the advancement of more people coming from the east and moving onto their lands, the Salish wanted the priests as allies and to strengthen themselves against enemies.

The Salish sent a group of their leaders east to St. Louis to meet with those who would listen four different times in 1831, 1835, 1837 and 1839.[6] Finally, during the fourth visit, the summer of 1839, Father Pierre-Jean De Smet agreed to meet with the delegation at what is today Council Bluffs, Iowa. De Smet accepted the invitation, and the Jesuit Society would set up a mission. De Smet, originally from Amsterdam, agreed to make the journey in 1840 to learn what would be needed.

On his return, De Smet organized and recruited a group to leave during spring 1841. Joining him were Father Gregory Mengarini and Father Nicholas Point. Three lay ministry brothers, four laborers and a guide who

St. Mary's Mission was founded in the Bitterroot Valley in 1841. Father Pierre Jean De Smet planted the first domestic garden and seeds of grains at the mission, the first introduction to agriculture. *Courtesy of the Ravalli County Museum Photo Archive.*

knew the area well also accompanied them. The party of eleven left on April 30, 1841, bound for the lands of the Salish. For a part of the trip, they journeyed with the first organized group of migrants traveling over the Oregon Trail to California.[7] On September 24, 1841, De Smet planted a cross, and Father Point journaled, "Everyone thought we would be able to find nothing better anywhere else. With one voice we said, 'It will be called St. Mary's.'"[8]

Introducing Agriculture

Construction began at once, and De Smet traveled to other mission stations in the Northwest to obtain seeds. The following spring, the first potatoes, wheat, oats and domesticated gardens were planted. The Salish thought it odd to destroy the grassland that fed their horses and a waste of food by putting it in the ground but were amazed when they saw the seeds sprout and grow. The fathers used this example to illustrate the fundamental mystery of Christianity, the resurrection from the dead.[9] The era marked the first principles of farming in Montana and the first lesson in raising food to eat. The first livestock arrived in the valley from Fort Colville. The brand used to identify the mission's cattle was the "cross on the hill." It is still registered to St. Mary's Mission and among the oldest official brands in Montana.[10]

In 1845, Father Anthony Ravalli arrived in the Bitterroot Valley after De Smet and Mengarini were reassigned. Ravalli, from Italy, was indefatigable and brought with him the skills, ingenuity and understanding needed for the mission to grow and survive. An educated, innovative man of industry and science, he created the items needed for church ceremony and to build the first gristmill in Montana with two twelve-inch small burh stones he had received from a local merchant in Antwerp. The mission's lay brothers, Joseph Specht and William Claussens, and French Canadian millwright Peter Biledot helped the priest develop the mission's flour milling operation and used Burnt Fort Creek to power it.[11] They also built the first sawmill, creating a saw blade from the metal of a wagon wheel.[12]

Despite the efforts of the priests, tensions from old feuds with tribal enemies, plus the pressure from outside western migration, began to reflect in the relationships with the Flatheads Indians. The mission's location was close to where the inland Mullan Road would soon be constructed. The route was already being used by frontiersmen, travelers and immigrants

Father Pierre Jean De Smet led the 1841 mission to the Bitterroot Valley at the request of the Salish Flathead Indians. The Jesuit priest traveled the Northwest Territories and was dedicated to his devotion in reaching out to the Plains Native American Indian tribes. *Courtesy of the Ravalli County Museum Photo Archive.*

Right: Major John Owen was the founder of Fort Owen in 1850 at the site of St. Mary's Mission. Owen hired men to raise grain at the fort for feed and trade. *Courtesy of the Ravalli County Museum Photo Archive.*

Below: Fort Owen was an important trading post in the Northwest and a major destination point for those traveling west. *Courtesy of the Ravalli County Museum Photo Archive.*

who arrived via the early pioneer trails or were traveling up the Missouri River to reach the river's far west settlement of Fort Benton in north-central Montana. The Mullan Road was finished in 1862, and supplies could then be freighted over land to and from Fort Benton to Fort Walla Walla, Washington.[13]

The customs and culture of the frontiersmen, the wagon trains and the early survey and road construction crews did not have a positive effect on the Native American Indians.[14] Father De Smet's missionary efforts in 1846 to reach out to the Blackfeet tribe were equivalent to treason to the Salish and Flathead Indians as it was believed De Smet was sharing sacred medicine with their mortal enemies. There was now an open hostility toward the "black robes" and indifference to Christianity.[15]

Relations worsened, and it was decided to close St. Mary's Mission after nearly a decade of service. On November 5, 1850, under the leadership of Father Gregory Mengarini, the priests made a conditional sales agreement, temporarily closing the doors.[16] The missionaries would travel north and continue their work at St. Ignatius in the Flathead Valley. The mission was dismantled and leased to Major John Owen. In 1866, Ravalli and Father Joseph Giorda returned to reopen the mission under a renewed commitment and interest from the Indians themselves.[17]

Major Owen did not hold an official military rank but earned and received his title as a civilian for his leadership and charge of the trading posts he developed on the frontier. Owen purchased land interests of St. Mary's and built a trading post close to where the original mission building was founded. He named the post Fort Owen after himself, and it became an important gathering place in the West.[18] Major Owen successfully raised crops and cattle to be sold and traded. A sawmill and gristmill served the inhabitants of the area for miles around. Thomas Harris, under Owen's employment, was the fort's farmer, and he is credited with the "first cropping" in Montana after raising a harvest of oats for the outpost in 1854.[19] Harris went on to become the state's first full-time farmer when he successfully started his own farm on land close to the fort. He planted his first crops in the spring of 1864.[20]

2

THE HUNT FOR GOLD FOSTERS AGRICULTURE GROWTH

Traces of gold began to appear in the Deer Lodge Valley in 1852. Francois Finlay, a prosperous trader in the Northwest, settled in the valley, bringing with him a drove of horses he had purchased in California.[21] Finlay, or Benetsee, as he was called, was of Scottish and Native American descent. He established his ranch before 1850 near a stream that became known as Benetsee Creek and based his travels from there. The stream and sandbars of the creek looked familiar to those he had seen on his far west trips to the gold fields. After acquiring a pan, he began washing the gravel, which produced about a teaspoon of yellow grains.[22] Finlay continued mining and prospecting for gold, but without any major success he returned to his frontier trading business.

News of Finlay's findings and rumors of gold reached those who were eager to leave the Colorado and California mining fields to explore new possibilities. Among those were the Stuart brothers, Granville and James. Originally from Virginia, the Stuart family moved to Illinois and then Iowa. Both young men had gone west to the Sacramento Valley with their father in 1852. After the patriarch of the family returned to Iowa, the brothers stayed in the California gold fields mining and herding stock. In 1857, they began their journey back home but were interrupted by bad weather, and Granville became sick with mountain fever. By the time he recovered, the Mormons had closed all the main roads, forcing Granville and James to accompany mountaineers who traded each summer with the migrants on the Overland Trail and wintered in the Beaverhead and Deer Lodge Valleys.[23]

By 1860, the brothers had moved to the Deer Lodge Valley and wintered at the mouth of Benetsee Creek, which they called Gold Creek. They intended to continue developing the gold mine interests in the vicinity and, by the following spring, had found several good prospects and immediately began mining operations. By 1862, others had arrived to work the gulches of the gold mining district. Some of the first gold seekers in the area were Conrad Kohrs, J.M. Bozeman, Nelson Story, Perry McAdow, Sam Hauser and Henry Sieben—all of whom, along with the Stuarts, played important roles in Montana's pioneer history, paving the way for others to follow.[24]

"The Rush Is On"

Gold! In July 1862, rich placer deposits of gold were discovered on Grasshopper Creek, a tributary of the Beaverhead River.[25] The Beaverhead Valley was 125 miles south of Gold Creek. Streams of immigration were on the way to Gold Creek and the Idaho mining fields. Once word spread of the Grasshopper Creek gold strike, hundreds immediately headed for the newly named town of Bannack. By the fall of 1862, a train had been sent to Salt Lake City for provisions, and by midwinter, 406 adult residents were living in Bannack—Gold Creek's numbers had fallen to 18.[26] The rush to Bannack was evident. By the summer of 1863, 3,000 to 5,000 miners, merchants, gamblers, saloon keepers, prostitutes and outlaws were living in Bannack. When Montana became a territory in 1864, the mining town became the first territorial capital.[27]

Another major strike in May 1863, seventy-five miles east of Bannack at Alder Gulch, set off a stampede of people rushing to the site. Within three months, a majority of the miners in Bannack had exited. The mining camp at Alder Gulch became Virginia City, a magic city of gold fields with a population that rose to more than twelve thousand by the end of 1864.[28] The territorial capital was moved to Virginia City, which soon became a prominent city of the Rocky Mountains. Many of Montana's "firsts" were recorded there: the first newspaper, the first public school and the first meeting of the Montana Historical Society.[29]

On July 14, 1864, another strike was found at Last Chance Gulch, 120 miles north of Virginia City. The new mining district town was named Helena. Again, thousands of miners, retailers, service providers and those seeking their fortune would head north.[30] Last Chance Gulch would prove to be a significant producer during the bonanza years of gold mining in the Montana Territory from 1862 to 1868.

Within six short years, the population of the Montana Territory had grown significantly. The boom of gold mining lasted into the early 1870s; by then, the towns of Gold Creek, Bannack and Virginia City had been nearly abandoned. Virginia City's population was only eight hundred in 1875.[31] Helena would survive the waves of mining, and the Last Chance Gulch strike brought economic development and stabilizing growth to the new city. It also helped that Helena was located on the main transportation routes, which provided a steady influx of agricultural products and other necessities.[32] The territorial capital was moved to Helena in 1875 and today is Montana's state capital.[33]

As Montana's "gold years" wound down, mining continued to play an important role in the state. Still to come were the developments of the ore-rich deposits of copper in the mining town of Butte. By 1876, it had been discerned that Butte was located on a hill of copper ore. Butte would be called the richest hill on earth, and soon one-third of all copper mined in the United States would come from Butte.[34] Thousands moved to Butte, Montana, to work in the great copper mines.

Feeding the Miners

Many of those arriving daily to the mining fields of the Rocky Mountains started their journeys in Europe or the eastern United States. They had been to the California gold fields and were heading north to seek their riches. Opportunity was waiting for them in the gold fields and in the businesses supporting the mining industry. The population boomed, creating glaring needs in the mining camps and towns for tools of the trade and food for survival. Being skilled at fulfilling these demands provided a ready-made job and means for a steady income.

Agriculture was pushed into production during the mining years, and the ability to produce food locally was realized out of necessity. Supplying products to the towns and camps springing up daily and delivering them safely would bring real wealth to those who were able to meet the challenge. Montana's early agriculture pioneers started their road to success with their skills, hard work, quick thinking and entrepreneurial spirit focused on feeding the miners.

The Miners Want Beef!

At age twenty-seven, Conrad Kohrs saw the Deer Lodge Valley for the first time in August 1862 after walking alongside the wagon train that brought him over one thousand miles. Kohrs came to America from Germany when he was fifteen years old. He worked as a butcher boy in New York City; a grocery clerk in Davenport, Iowa; a lumberman in Wisconsin; and gold miner in California.[35] At Gold Creek, he met the Stuart brothers, who had settled the mining district town and were instrumental in helping organize the city of Deer Lodge, formerly called Cottonwood. Kohrs quickly became acquainted with others who came by wagon to work the diggings and those who traveled west by steamboat on the Missouri River to Fort Benton and then used the new Mullan Road to get to the mining fields.

Word of the Grasshopper Creek strike came quickly, and immediately Kohrs joined the others who headed south to Bannack. Along the way, they met Henry Crawford, who ran a supply line to the mining camps. After giving a brief update, Crawford asked if there was a butcher in the crowd. Kohrs knew the only way to survive was to take some type of service job in the mining camp. With a few simple tools, he went to work and was off to a solid start. Crawford left Kohrs in full charge of the butcher shop, purchasing livestock, processing and selling the end product to the miners.[36] Crawford was forced to abandon his supply line and businesses after run-ins with the "road agents" or "highwaymen," notorious robbers and killers roaming the hills and roads to and from the mining camps. Before escaping, he turned his supply line over to Kohrs, who was now in full-time business for himself.[37]

As a new owner, Kohrs needed an inventory of beef and relied on those he met when first arriving by wagon train. His new acquaintances had been raising herds of cattle on the lush grasses of Deer Lodge Valley. As he had no money, Kohrs made arrangements to purchase livestock on credit for the mining towns, which included the camps at Alder Gulch and the new town of Virginia City.[38] The demand for livestock escalated, and Conrad Kohrs had the energy, skill and business acumen to find and secure the supplies that were needed.

Kohrs constantly searched for livestock. Johnny Grant, a Canadian trapper and mountaineer, settled in Deer Lodge Valley. By 1850, Johnny, with his brother James and their fur-trading father, Richard Grant, had begun trading with the travelers on the pioneer overland trails. When the wagon trains reached southwestern Montana and southeastern Idaho, their animals were worn out and undernourished. After trading for livestock, the

Grants trailed the animals to Beaverhead Valley to winter on the tall grass. The following spring, they moved the fattened cattle back to the Oregon Trail, trading one fresh animal for two that were trail weary.[39] In less than a decade, Johnny Grant had more than two thousand head of cattle grazing in the Deer Lodge Valley. The large herd marked the beginning of the range cattle business in the Montana Territory.[40] In 1862, Grant built a four-thousand-square-foot house at Cottonwood for his family, the largest home in the territory.[41] It often served as the hospitality and gathering center for the valley.

Conrad Kohrs regularly purchased cattle from Johnny Grant, who made practical loan arrangements for the young beef buyer. The cattle owner kept no books and had built a prosperous relationship with Kohrs after witnessing that he reliably paid his bills and with interest.[42] Kohrs discovered the "true gold" was in providing the meat supply for the butchers in the mining districts from Virginia City to Helena and from Butte to Deer Lodge. He was beginning to provide meat for the soldiers at the various and new military posts being built. In 1865, Kohrs purchased Grant's ranch at Cottonwood and based his operations in the Deer Lodge Valley, where he lived the rest of his life and raised a family.[43]

Kohrs became Montana's leading stockman and continued to develop the state's livestock industry; he was the first to ship cattle to national and international markets. During the 1860s and 1870s, Kohrs was dedicated to improving his herds by replacing the Texas longhorn cattle with shorthorn and Hereford breeds. Over time, the herd grew too large to winter in the Deer Lodge Valley. In the fall of 1867, he sent close to one thousand head of his best cattle to the grasslands south of the Sun River, marking the initial entry of the first hundreds of thousands of cattle that would eventually cover the Montana plains—the last free-grass area in the nation.[44]

The Grant-Kohrs original ranch remains a year-round working ranch and a National Historic Site preserved in its original state by the family's efforts and the National Park Service. Visitors are welcome to tour and step back in time to experience Montana's first major cattle-working operation. The legacy of Conrad Kohrs, one of the state's earliest pioneers who became Montana's Cattle King, is shared at the visitor center and through tours at the historic ranch site located at Deer Lodge. More information is available at www.nps.gov/grko.

The Bozeman Trail and the Gallatin Valley

John Bozeman came from Georgia to work in the Bannack and Alder Gulch mines. By early 1863, Bozeman was convinced he could find a faster route to the gold fields from the Oregon Trail following the North Platte River. With his business partner, John Jacobs, Bozeman scouted a new shortcut.[45] It became known as the Bozeman Trail and was an alternate passage connecting with the Oregon Trail fifty miles northwest of Fort Laramie, Wyoming. Those who ventured along the trail could immediately turn north and travel the eastern side of Wyoming's Big Horn Mountains to the Yellowstone River in southeastern Montana, then cross the Bozeman Pass into the Gallatin Valley.[46] From there, it was a day's ride to the gold fields.

Once they reached the Gallatin Valley, the migrant wagon trains traveled through a settlement called Jacob's Crossing, the first supply outpost on the trail. John Bozeman picked the spot for the outpost, knowing it looked promising for a future agricultural town. On August 9, 1864, at a meeting of the Upper East Gallatin Association, Jacob's Crossing was renamed Bozeman.[47]

The newly founded community was "being fast settled up with farmers," according to John Bozeman. By 1865, about 20,000 bushels of wheat were produced on about 1,500 tilled acres in the Gallatin Valley. One of the first pioneers to grow wheat in the valley was John Thomas, stepfather of Henry Davis, who brought a bushel of wheat with him from Utah. After planting it in the spring of 1864, Thomas reaped 50 bushels from his crop.[48] He threshed the wheat by laying the stalks with the heads of wheat pointing out and having horses walk over it until the grain was separated from the chaff. Thomas and Davis would save the harvest and sell it as seed to farmers the next spring for ten dollars a bushel.[49]

John Bozeman met Perry McAdow and his brother, William, after arriving in Gold Creek to begin mining. He encouraged the McAdows and Thomas Cover to start a flour mill together. By 1865, their new mill was the first commercial operations in the area and produced one thousand sacks of flour per week.[50] The two millstones for the enterprise, each weighing one thousand pounds or more, were purchased in St. Louis and brought to Fort Benton by steamboat and transported to the site on wagons pulled by teams of mules.[51] The millstones are on display in the southeast corner of Bozeman's Beall Park.

In retrospect, it was written that President Thomas Jefferson, known as one of America's first agrarians, "would have found his dream of an agrarian-based America alive in the early Gallatin Valley. By the mid-1860's, the first crops were harvested with wheat, potatoes, grains, and vegetables finding markets in regional mining communities."[52]

Even with the rich soil and an abundant water supply, the settlers were forced to deal with the region's unpredictable weather, which produced shorter growing seasons with hard freezes and early snowfall. In 1883, Matthew Alderson, who settled near Bozeman with his father, commented in his diary, "A man to withstand the rigors of any northern climate must have at least the courage at heart to brave almost anything and fight his way through."[53]

The First Texas Cattle Drive

When Nelson Story and his wife, Ellen, arrived in Bannack in 1863, only women and children were there. The men were gone, having rushed in stampede fashion to the Alder Gulch strike discovered a week earlier. Story immediately left for Alder Gulch and was able to stake his claim. He brought ox teams and fourteen pack mules with him and made money hauling for the miners, while Ellen baked bread and made pies to sell.[54] By 1866, their claim had paid off, producing $30,000 in gold. Story immediately took the capital he earned and used it to build a future for his family in the Gallatin Valley.

In a special collection of interviews for Montana State University, Nelson Story's grandson Malcolm Story shared the following insights:

> *He hid his treasure in a tin box, strapped it around him and headed back East, leaving Ellen with a Bozeman preacher and his wife. For the gold, Story got $40,000 in greenbacks, because gold was at a premium after the Civil War. He put $30,000 in a bank and sewed $10,000 inside of his overcoat. Story and two trusted friends headed for Texas. At Fort Worth, he bought 1,000 head of longhorn cows for $10 a head.*[55]

For the next nine months, Story and his crew of hired cowboys brought the longhorns north to the upper Yellowstone Valley, near Livingston, marking the first time eastern Montana entered the economic picture.[56]

This was the first of the cattle drives from Texas to the Montana Territory over the next three and a half decades to take advantage of the free grass of the northern plains. The drive was the inspiration for author Larry

McMurtry's Pulitzer Prize–winning novel *Lonesome Dove*.[57] Story ranged his cattle east of Bozeman in the Paradise Valley, growing his herd to fifteen thousand by 1886.

The Bozeman entrepreneur continued to help develop the economic growth of the Gallatin Valley, including the dairy and grain industries. In 1882, he built the Story Flour Mill, which produced one thousand bushels a day. The mill survived a fire in 1901 and was rebuilt by 1904 with improved facilities and "captured the market for central Montana wheat milling and storage."[58]

An important part of Nelson Story's legacy is his contribution to the establishment of the Agricultural College of the State of Montana. As a member of the State Board of Education, his lobbying efforts to the state legislature were instrumental in the decision in 1892 to locate the college and the new State Agricultural Experiment Station in Bozeman.[59] He helped launch the school by donating 160 acres of land for the site, allowing the school to open by July 1, 1893, and winning $33,000 from the federal government for meeting the deadline.[60] Today, the college is known as Montana State University.

3
PIONEERING RESPONSIBLE CITIZENSHIP

Henry Sieben was a young pioneer who traveled to the gold fields by way of the Bozeman Trail. He was the youngest of six children and at the age of seven emigrated with his family from Abenheim, Hesse-Darmstadt, Germany, to the United States, where the family eventually settled in Geneseo, Illinois.[61] Ten years later, after losing both of their parents, Henry and his brother Leonard headed to the Alder Gulch mining fields in Montana in April 1864. They each paid ninety dollars to young wagon owners Louis Heller and Philip Arnett, who acquired the wagon from their fathers. The passage included food, and the four took turns driving the wagon and walking.[62]

When they reached Fort Laramie, Wyoming, the military refused the young men permission to continue on the Oregon Trail, so they decided to join a train of one hundred wagons being led north by John Bozeman on his newly scouted trail.[63] They arrived in the western Montana mining camps in July. Henry Sieben immediately took a job on a ranch cutting hay in Madison Valley on Meadow Creek east of Alder Gulch. In the spring of 1865, he took his earnings, walked forty miles to the Gallatin Valley and purchased an oxen team to take to Virginia City to begin a freighting business.[64]

Henry and Leonard Sieben worked together in the freighting business through the summer of 1870, when the brothers sold their operation and equipment to the well-established Diamond R Freight Company. The two already had a few cattle and would purchase more to begin ranching at their new Smith River headquarters with a herd of 400. The Siebens' brother

Jacob, from Iowa, would later join their partnership and was interested in raising sheep. Henry and Leonard furnished the money for Jacob to buy 2,200 head of Merino sheep in California and trail them back to Montana in 1875, marking their initial entry as one of the earlier pioneers into the sheep business.[65] In 1879, Leonard Sieben sold his ranching interests to Henry and returned to Illinois to establish a farm. At this point, Henry took over the cattle and horse interests while Jacob looked over the sheep.[66]

A leading figure in the livestock industry by 1880, Henry Sieben was well established before the cattle boom involved the eastern side of the state. He had worked in the ranching business since arriving in the territory, and in 1893, Sieben purchased several small ranches near Culbertson with Helena business partner Elizur Beach to create the Diamond Ranch, which became one of the largest cattle ranches in eastern Montana. Frank G. Arnett, Sieben's nephew, joined the two and oversaw the activities on the ranch. Later, Arnett took over the Diamond operation when Sieben purchased the Mitchell Ranch north of Helena to base his operations. In 1907, Sieben was able to buy the Adel Ranch above Cascade near Great Falls. The Adel was a ranch he had admired after observing a certain spot where the cattle always drifted, commenting to his brothers "that a good ranch to buy is a place where the cattle chose for themselves."[67]

By 1907, the Mitchell and Adel Ranches had become the heart of Henry Sieben's cattle and sheep raising business—later organized as the Sieben Livestock Company with the Adel Ranch operating a large sheep business. For over a quarter of a century, the Sieben Livestock Company was one of the largest sheep concerns in the state.[68]

Henry Sieben was good at the business of livestock and helped shape the cattle and sheep industries of Montana. Establishing a reputation as an excellent businessman, he took care of his employees and would not stand for the mistreatment of stock. Sieben was progressive and well thought of by his contemporaries. Dick Pace, who wrote the article "Henry Sieben: Pioneer Montana Stockman" for the Montana Historical Society, shared these thoughts after doing extensive research: "The most remarkable thing about Henry Sieben, is that he seems to have left this world with no enemies—he was a loved and respected man. And that I think is most unusual."[69]

Today, the Sieben Livestock Company at Adel, Montana, and the Sieben Ranch Company, twenty miles north of Helena, are still owned and operated by family.[70] Henry's great-grandson Max Sieben Baucus is the longest-serving U.S. senator for Montana, elected to office for six terms from

1978 to 2014. Senator Baucus represented America as the U.S. ambassador to China from 2014 to 2017.[71]

Production Agriculture in Full Swing

Growing food was well underway in the new Montana Territory by 1866. With an emergent population to feed, many who came searching for gold had by now committed themselves to farming and ranching ventures. Experiments with growing different types of food were conducted with great interest. Attention was on viable products that could survive the climate and weather to successfully make it to the markets in time to realize a profit.

Enough food was being grown by the new agrarians to feed themselves and their families and establish trade with the mining fields, military forts, immigrants and construction crews building roads and the future railroad. Demand for food was beginning to be filled by locally grown and milled commodities. Although production had increased, problems of limited access to markets and getting products to consumers from remote locations were still challenges to overcome.[72]

As Bitterroot Valley grew, others soon followed Thomas Harris in developing their own farms, primarily producing wheat, oats and barley. Father Ravalli planted fruit trees at St. Mary's Mission when he arrived in 1845. In 1866, the priest returned to reopen the mission, and the apple trees he planted had survived. Harris also planted fruit trees on his farm and is credited with planting the first orchard in 1866. From a Washington-based salesman, Thomas purchased one hundred apple, four plum and two pear trees, as well as twelve raspberry and fifty strawberry plants for his first orchard planting.[73]

4
MONTANA'S EARLY AGRICULTURE BLOOMS AND BOOMS

The Apple Boom

Every year, the first Saturday of October, heralds an important time of history for the town of Hamilton and the entire Bitterroot Valley. Apple Day is a signature event created by the Ravalli County Museum to commemorate the apple boom of the valley during the early 1900s.

While others laughed and scoffed at the idea of raising apples commercially in the valley, there were those who were busy planting trees. In 1865, the Bass brothers, Dudley and Edward, arrived in Stevensville, the new town next to St. Mary's Mission and Fort Owen. Five years later, in 1870, the duo challenged the critics and planted the first successful commercial orchard on the west-side bench lands and called it Pine Grove Fruit Farm, a location well suited for fruit growing. The Basses also opened an experimental station that operated for twenty years. They planted a variety of trees and in 1877 harvested their first apples, primarily McIntosh.[74]

Apple production grew in the valley and by 1880 had reached one thousand bushels. Stevensville held a Fruit Fair in 1883 to celebrate the success of the apple industry.[75] Pine Grove Fruit Farm was well established by 1890, and the Basses shipped apples to Butte. The orchard brought in $1,000 that year.

The opening of the Northern Pacific Railway in Montana in 1883 and the construction of the Missoula and Bitterroot Railroad Company, a branch

Creating Montana's Agriculture

A bountiful harvest. Crates of apples ready for shipping from the Bitterroot Valley's apple harvest in 1910. Ravalli County historian William W. Whitfield noted by 1915 the local orchards were yielding up to 230 boxes per acre at $2.50 a box. *Courtesy of the Ravalli County Museum Photo Archive.*

line of the Northern Pacific, brought farmers to the valley and facilitated access to markets for their crops. Apple production was on the rise, and by 1896 the Bitter Root Orchard Company had the largest orchard in the world, with 40,000 trees on 380 acres. By 1907, valley growers shipped 250 box car loads of apples out in 150,000 wooden boxes. The apple boom was underway with irrigation ditches and canals being initiated and completed. The year 1921 was the peak of the boom, and growers loaded 637 boxcars with apples from the harvest to be shipped out by rail.[76]

Unfortunately, the boom that started in 1910 after the main irrigation ditch was completed dwindled as fast as it came. The weather would destroy the orchards over the next three years. Hail decimated the crops and most of the orchards in 1922 and 1923 with a killing frost the following year, freezing all the buds on the trees.[77] The orchards could not recover from the three years of bad weather, and the valley's promising future in the apple and fruit industry was over.

Apple pickers harvesting apples from a Bitterroot Valley orchard. The valley is well remembered for growing MacIntosh apples, known for their tartness and sought after for baking pies. Dubbed "Montana Macs," the apples were marketed and shipped throughout the nation. *Courtesy of the Ravalli County Museum Photo Archive.*

The Cattle Boom

As Montana's gold mines and strikes ran their course in history, another treasure appeared on the territory's eastern plains. Men who had struck it rich during the mining years were now eager to invest in cattle. Investors in the eastern United States and Europe were ready to partner and create alliances with the cattlemen in the West. The western valleys became overgrazed, herds continued to grow and the search for new pastures was ongoing.

By 1880, Montana had experienced a major transition in the name of western expansion, dramatically changing its culture. The buffalo that once roamed the high plains were gone, and the Native American tribes had been removed from their prairie homelands to reservations. A sacrifice to pay for development, and now, two-thirds of the state was open for business and the grass was green.

Large cattle companies immediately formed, quickly filling up Montana's free range from the Wyoming border to the Dakota and Canadian lines. One of the largest cattle partnerships was the DHS Ranch, operating in the

Judith Basin. Sam Hauser, who came to Gold Creek in 1862, and Andrew J. Davis and his brother Erwin, along with Granville Stuart, who had arrived in the territory in 1857, formed Davis, Hauser and Company in August 1879. Andrew Davis arrived in Montana and made a considerable fortune in mining by 1870.[78] Stuart was named general manager and superintendent of the company. His first task was finding suitable rangeland and stocking it with cattle. Granville chose the Judith Basin and selected a site on Ford's Creek to build the ranch headquarters. Stuart, with his family, would live year-round at the site.[79]

Granville Stuart was actively involved in the range years from 1880 to 1887. Within two to three years, the cattle flowed into the grasslands, causing the inevitable need for an organized, self-regulating group.

> *In 1883 to 1884 the Montana cattle industry moved rapidly toward a long-needed, self-regulating organization. Beginning with the Shonkin district in 1881, local cattlemen's associations had sprung up in several areas; then, in Miles City on October 12, 1883, a group organized the first major association: The Eastern Montana Stock Growers Association. The next spring, at a meeting in Helena, Conrad Kohrs and Granville Stuart led in the formation of the Montana Stock Growers Association, of which Stuart served as President. In agreement on principles, the two associations merged in April 1885, at a joint meeting in Miles City. The president of the eastern association became president of the new association, and Stuart moved to the executive committee. In the same year the association issued its first brand book.*[80]

Overstocking and over-grazing emerged. Cattle disease, cattle rustling and grassland depletion ensued as a result of free range. The U.S. Department of Agriculture estimated that in 1883 there were nearly 600,000 cattle and 500,000 sheep in the territory and most of them east of the mountains, with the center being in Custer County, where Miles City serves today as the county seat.[81]

Most of the larger cattle companies were capitalized by outside money, but there were Montana-owned enterprises among the largest outfits. The DHS had twelve thousand head of cattle by 1883. In 1885, Conrad Kohrs bought a large interest of the DHS Ranch for $1 million, creating the Pioneer Cattle Company, the largest cattle company in Montana.[82]

The Blizzard of 1886–87

The summer of 1886 brought drought conditions, limiting grass availability, and by that fall, cattle were in poor condition. Owners worked to sell off as many head as possible, but prices were the lowest they had been in years. Granville Stuart got caught in the first blizzard of the winter while doing an inspection north of the Missouri River. He remarked that the storm, which started on Christmas Day, continued to blow with varying intensity for the next two months: "It was as though the Artic regions had pushed down and enveloped around us."[83]

The weather took its toll on the cattle. With no grass under the snow for food and much of the original shelter bottoms being fenced, the great cattle die-off began. The losses were catastrophic, with some cattle producers losing more than 75 percent of their herds. Many called it quits. The disastrous winter marked the beginning of effective range management and a new era for the cattle industry. Once again, the weather shut down an agriculture economic boom and changed the state's history. Granville Stuart echoed the sentiments of the cattlemen and cowboys who rode the range to survey the blizzard's destruction: "I never wanted to own again an animal that I could not feed or shelter."[84]

5
THEY CAME WEST TO BUILD A NEW FUTURE

THE KENTS

With rumors of "new strikes," the migration of people to Montana continued at a steady pace—all coming to the gold fields from the eastern United States, torn by the Civil War, Europe, Russia and the far eastern shores of China. The ore-rich valleys and gulches also drew them by the hundreds each day. All attention was needed to navigate their new surroundings, to find work, safety and shelter and to take advantage of the opportunity they had persevered and strived so hard to get. Now was their chance to discover if they could, in fact, make it.

Thomas Kent

The main street in Virginia City was hot and dusty when twenty-two-year-old Thomas Kent arrived with his freight wagon full of goods to sell on July 27, 1864. His 3,200 pounds of freight consisted mainly of whiskey. The weight-bearing load met the military's restriction of 800 pounds to be pulled per animal. Kent knew how to freight. He made a living trading and freighting in the Nebraska Territory with his favorite four-horse team. His plans now were to make enough from the sale of the whiskey to support his efforts to mine for gold. After meeting James Gamble, he immediately went to work freighting and working on Gamble's farm on the Ruby River. After six months, Kent returned to Alder Gulch, where he actively engaged in mining.[85]

Born in Mount Pleasant, Pennsylvania, to Samuel and Mary Kent on January 3, 1842, Thomas was one of eleven children. Young Kent moved with the family to Des Moines County, Iowa, when he was ten. The Kent family operated a successful farm and livestock sales barn selling horses, cattle and sheep. He preferred to be called "Tom" and was an accomplished horseman who had his four lead horses with him all the time.[86]

Tom Kent's journey west started on March 24, 1864, from eastern Nebraska and the North Platte River. Fifty miles past Fort Laramie, he met John Bozeman and befriended the wagon master, who was preparing to lead the 125-wagon train Kent had joined. The wagons would travel north on Bozeman's newly scouted trail. Bozeman was congenial, tall, strong and not ruled by fear. Those who knew the leader said he could carry a load that would break a mule.[87] Before the wagon train left camp on the first day to travel north, it was recorded that Bozeman summed up the journey ahead: "I want none but brave men, as we are going through a hard country and cowards had better stay back."[88] On July 4, the wagon train was close to present-day Livingston, ironically near the same area Kent would chose to build a ranch, farm and home starting in 1878. At that moment, Kent was only twenty-three days away from his destination of Virginia City.

Kent would make another trip up from the North Platte River to Alder Gulch, perhaps a freighting trip or coming back from seeing family in Iowa. This time he traveled back to Montana on the Bridger Trail, led by the trailblazer himself, Jim Bridger. The trail went to the west of the Bighorn Mountains and was regarded safer.[89]

> *Authors' Note: Stan Stevens, Tom Kent's great-grandson, confirmed what the history books revealed about his relative's early contributions in settling Montana and contributing to the state's early growth. At ninety-two, Stevens, with a good nature and clear memory, shared stories about the early pioneer ventures of Kent during an oral history interview in October 2016. Over his lifetime, he had listened to detailed accounts told to him by his elders and his grandmother Mary Stevens, Kent's daughter. Stan's relative, Leonard Dailey, Jr., a great-great-grandson of Kent, took part in the interview to help identify locations, timelines and clarify family connections. Dailey and his wife, Kathy, were readily available to answer questions and share information while researching the history and legacy of Thomas Kent.*

In the spring of 1865, Kent returned to Virginia City. After working and freighting for James Gamble, he was ready to start mining for gold

Buffalo Run captures the herd moving across the High Plains. *Courtesy of Gene Roncka, Willow Point Gallery, Ashland, Nebraska.*

and bought a placer claim for only $400. The fact that Kent was able to purchase the claim for such a low price and with an "over-time" purchase plan indicates the seller did not think the ground was worth anything and believed he had taken in a poor tenderfoot with no clue of the work involved to work the diggings.[90] Within four months of taking ownership of the claim, Tom had gleaned $15,000 in gold.[91] By the time Kent finished working the claim, he had earned $27,000. Being known for possessing a "nature ever proverbial for its generosity and open-heartedness, Mr. Kent then made a present of the claim to a friend who had been less fortunate."[92]

Continuing to prospect the new strikes at gulches in the gold fields, Kent purchased his second and last claim in the Blackfoot Country at Lincoln Gulch. By the time he reached bedrock, he had discovered the mine was barren, and with his time, energy and money spent, Kent left the mining business.[93]

Tom spent a season traveling the Yellowstone Valley, trapping and participating in "wolfing" on Flat Willow Creek. He was an excellent hunter and yielded over eight hundred wolf pelts to sell. The time on the Yellowstone would expose him to the abundant grass and range lands from the Bozeman Pass to the Yellowstone, Big Horn and Powder River Country. Within ten years, Tom Kent would be one of the leading livestock pioneers in this important region of the territory.

Times Changed

It was now 1869, less than thirty years since Father De Smet arrived in the Bitterroot Valley on the invitation of the Salish Indians and less than twenty years since the Fort Laramie Treaty of 1851 (Horse Creek Treaty) had been signed at the mouth of Horse Creek on the North Platte River, between present-day Morrill and Henry, Nebraska. More than ten thousand Plains

Indians (men, women and children) with their tribal leaders gathered on September 1, 1851, for the great treaty council on the invitation of the U.S. government.[94] The location was Fort Laramie in the eastern part of what would become the state of Wyoming in 1890. Strategically positioned on the Oregon Trail, it was one of the most important forts built in the west.

Unable to accommodate the size of the crowd at the fort, the council was moved thirty miles east. Father De Smet was among those in attendance and was welcomed by the Indians and the government officials as an emissary of peace. Never before had so many American Indians assembled of their own will to negotiate and talk to the white man. This council is viewed as possibly the "most dramatic demonstration of the Plains tribes desire to live in peace."[95]

The purpose of the meeting and the intent of the treaty were to outline the rights and responsibilities of both the American Indians and the U.S. government, as the "encroaching" European American population was competing with the tribes for available resources.[96] There were tensions on both sides as the group parlayed for over seventeen days, working to create a peaceful resolve between both groups. The U.S. government also wanted to establish the right to make roads and establish military and other posts within the Indian Territories. In addition, the U.S. representatives agreed to protect the Indians from depredations of the white settlers and provide annual compensation to the tribes. Each tribe was to pick a "principal chief" to represent their interests, and for the first time, boundaries were discussed and established for the individual Indian nations."[97]

On September 17, the Fort Laramie Treaty of 1851, known as the Horse Creek Treaty, was signed. Those signing the treaty were representatives of the U.S. government, David D. Mitchell and Thomas Fitzpatrick, appointed and authorized by the president of the United States Millard Fillmore, and twenty-one chiefs for the Indian Nations, each signing with an X and having a signed witness.[98]

The Horse Creek Treaty with its preamble and articles created on paper what could only be wished for later. What ensued for years to come were battles of embitterment against both sides who signed the treaty and between Indian nations themselves. Broken promises and failed treaty agreements continued. Destruction of life, land and principle escalated with the pressures to develop lands for grazing purposes, cultivation of the prairies and harvesting the rich resources in the mountainous regions. It was the end of the hunting grounds and homelands of the American Plains Indians and the beginning of Native Americans living on designated boundaries of land known as Indian reservations.

Butcher at Flesher Rock

For the Crow Nation, the transition started in 1870, when the Native Americans were expected to move onto the territory defined by the Fort Laramie Treaty of 1868, but their territory boundaries had been reduced since the first treaty of 1851.[99] A government agency was to be established on each reservation to handle the affairs of the tribe and for the distribution of goods, foods, and services agreed to by the U.S. government in exchange for Indian lands.

The Mission Creek Agency was founded on Mission Creek near the Yellowstone River east of present-day Livingston. The Crows called it Flesher Rock Agency since it was located across the river from a familiar geological landmark: a tall bluff overlooking the river with the top surface made of flat rock. Flesher Rock was used by tribes to skin, dress and dry the hides of animals from their hunts.[100]

The government needed someone to handle the livestock allotted to the tribe for food. Established cattle ranches like those of Conrad Kohrs of Deer Lodge and Nelson Story from Bozeman were some of those who negotiated contracts to supply beef for the reservations. Tom Kent worked for the U.S. government for several years managing the herds for the Crow tribe and handling the processing and butchering.

During this time, Tom met and fell in love with What She Has Is Well Known, later called White Shield, the granddaughter of Crow Indian chief Old White Head. Tom paid the honorable price for the right to marry White Shield, consisting of five fine horses, guns, ammunition, blankets and other gifts, and it was seen as a mark of distinction.[101] The two were married in a traditional Indian ceremony, and she became known as Mary Kent. Later, they and eight other couples were married by a Catholic priest. Elizabeth McComas, a younger daughter of Tom and Mary Kent, recorded her parents' wedding date as April 10, 1874.[102] The couple had five daughters: Ella, Mary, Josephine, Elizabeth and Margaret.

Tom Kent assisted in moving the Crow beef herd and meat processing operation in 1875 to the new Crow Agency by the Rosebud River, near the present-day town of Absarokee.[103] While Tom herded the cattle to the agency, Mary handled getting their personal belongings moved by an ox-drawn wagon.

After the agency was moved to Rosebud, Tom left his position with the government. He began a large trading business and soon was able to build his own ranch. According to P.J. Smith, a Kent family relative and researcher with the Yellowstone Genealogy Forum in Billings, Tom Kent started the ranch in 1878 on his wife Mary's allotment near Greycliff by Bridger Creek.[104] Ella Cashen, the Kent's eldest daughter, confirmed the location and timeline when interviewed during a visit to the old Mission Creek Agency in 1960 with younger sister Josie Bostwick.[105]

Bringing the Sheep Industry to Big Timber

The Kents' ranch was located in the first areas of the Yellowstone Valley that Tom saw while traveling on the Bozeman Trail to the gold fields in 1864. In addition to his wife's allotment, Kent purchased nearly one thousand acres at Greycliff and raised cattle and horses until selling off the livestock in 1883 to go into the sheep business. He would be one of the first in the sheep business to successfully enter the territory's new industry.

> *John Spomer, great-great grandson of Tom and Mary Kent and great-grandson of their daughter Ella Cashen, shared an interview at Bighorn County Museum in Hardin with author Melody Dobson about Kent's entry into the sheep business from stories handed down to him by family and through his years of research. Spomer noted, according to the* Billings Post, *that Kent sold 1,200 head of cattle in the spring off 1883 to Briggs & Ellis out of New York and Tom trailed them to Stinking Water or Ruby River.*[106] *Now he was free to make preparations for bringing sheep to the Bridger Creek Ranch area.*

Before Kent traveled to purchase the sheep, he wanted to visit his mother in Iowa. Federal authorities were suspicious that he was planning to leave his Indian family like other white men in the territory had done and seized his monies and did not release the funds until he could prove that he was coming back.[107] Tom did return and brought with him his widowed mother, who lived with Kent's family until she died in 1891. Mary Kent, with the help of Tom's mother, would look after the affairs of the ranch and raise the family while Tom was away driving the sheep back from Oregon, a venture that took over a year.

In 1884, Kent acquired close to four thousand head of sheep from the sheep country of the John Day River Basin in central Oregon. Herding the sheep to the ranch would take in account the time needed for shearing, selling the wool, lambing and waiting for swollen rivers to return to their banks and run lower to allow for safe crossings. Kent was photographed bringing bands of sheep over one of the first bridges built in the Gallatin Valley.[108] A band of sheep is equivalent to one thousand ewes and with newborn lambs would equal close to two thousand animals.

A Ranching Legacy for Generations to Come

Kent is credited for bringing sheep to the Big Timber region and developing it into one of the major centers for the industry during the late 1880s. He had as many at twenty-five thousand sheep at one time. Kent's sheep could be seen from Big Timber to Miles City and at Lodge Grass on the Crow Indian Reservation. Sheep shearing sheds and lambing barns were at different locations and provided local jobs. With the Northern Pacific Railroad coming through in 1883, it was much easier to bring buyers in and get wool shipped to the markets. A local landmark's name was changed to reflect the growth during that period: Flesher Rock was renamed Sheep Mountain after a band of sheep were driven over the rim to their death by a howling blizzard.[109]

Tom Kent survived the volatile markets of the sheep industry and continued to help people get started in the sheep and the livestock businesses. He was progressive and forward thinking. From his account ledger entries, payments were made to those who helped him with surveying to build the Kent Irrigation Ditch, bringing water for hayfields and the other new crops being introduced.[110] Spomer shared that his great-great-grandfather was one of the first in the area to have a tractor to plow and plant the land.

Mary and Tom, both well-liked and respected, built the first frame dwelling in the valley. The home was a landmark and built with most of the lumber freighted from Bozeman by teams. Construction on the two-story home took place from 1884 to 1888.[111] The *Big Timber Pioneer* newspaper stated in 1903,

> *Besides being the oldest pioneer of the Yellowstone who has made a part of that wonderful and historic valley his home, having been identified in paving the way for its civilization, at the early year of 1864, Mr. Kent also retains the distinction of having the finest ranch residence, as well as the best and costliest furnishings of any throughout the length and breadth of its domain; furthermore, his home residence, is the first frame dwelling that ever graced the valley. It is made the more homelike and comfortable*

by its orchards of apples, plums, cherries and all varieties of small fruits, while to his comfort added an amiable wife and helpmate and five children. Possessing the above noteworthy attainments, the Kent home is the only true relic of pioneer days in which the "latch string is always on the outside," for, like those good old frontier days, none were barred from entrance to their door, consequently this venerable couple today recognized no enemies, treating alike rich and poor, high and low—it matters not who.[112]

Through the work of his sons-in-law and daughters, Tom Kent extended his ranch holdings and efforts through south-central Montana and onto the Crow Reservation, where he ran cattle, sheep and horses. When Tom passed away at the age of sixty-eight on August 17, 1910, he left the ranches to his fifty-year-old wife, Mary (White Shield) Kent, who operated the ranching business, which had grown by then to several thousands of acres. She attended to the day-to-day affairs up until her death on May 3, 1919. Her family praised her for the great job she did and was honored by her Native American heritage as a member of the Crow Indian Nation.

The ranching legacies and traditions of Tom and Mary Kent continue well into the third, fourth and fifth generations. In 1989, another great-grandson of the pioneer couple, Jay Stovall, was named the trail boss for the organized Centennial Cattle Drive honoring the one-hundredth anniversary of Montana's statehood.[113] Stovall served as a legislator in the Montana House of Representatives for eight years and for four years as a Montana public service commissioner. As an enrolled member of the Crow tribe, he was dedicated to working hard to make sure the tribe's voice was heard. During President George W. Bush's years in office, Stovall was appointed to represent Indian and minority issues in education and the U.S. Department of Agriculture.[114]

Authors' Note: Jay Stovall's son, Turk Stovall, is one of the younger generations continuing the Kent ranching legacy. In an interview with author Melody Dobson, Turk praised the legacy of his pioneers and family members who have worked hard their whole lives in the livestock industry. As Turk and his wife, Jenny, continue to embark in their own ranching endeavors east of Billings, he gave tribute to his parents, sharing his Dad was an "Oak." With a solid background to reference and gain knowledge from, the younger Stovall described the task at hand today in ranching as managing risk with diversification. Easier said than done, but daily site involvement helps keep the margins positive.

Seated in the center of the photo are Mary (White Shield) Kent and Tom Kent before 1910. Tom's sister Mary Jane is standing on the left, and his sister Kate is on the right. *Courtesy of descendants of Thomas Kent and Mary Kent (aka White Shield aka What She Has Is Well Known).*

The two-story house of Mary and Tom Kent, located at Bridger Creek near Greycliff. The beautiful home could be seen for miles and was the first framed house constructed in the area. *Courtesy of descendants of Thomas Kent and Mary Kent (aka White Shield aka What She Has Is Well Known).*

During the interview, Turk gave a driving tour of nearby pastures as he investigated the conditions and pointed out the different grasses and food sources the cattle depend on during an open winter. He also checked on the access road he was having carved out and built to ensure feed could be delivered to the livestock during winter months. Stovall gave a brief tour of his cattle finishing operation. He described the work that has been done to update the watering system and install state-of-the-art equipment to adequately dispense the right nutrients and feed for proper nutrition. All the steps are monitored and maintained for feeding at the appropriate times. The special tagging and penning system lets Stovall know at what stage his feeder stock are in at any given moment.

Resourceful and progressive just like his great-great grandfather Tom Kent, Turk Stovall displays a natural enthusiasm and energy for the career he has chosen and one that is graced with admiration for his Kent legacy and pride for his White Shield heritage. After visiting about the frustrations and challenges facing the cattle industry, it is easy to recognize the same adventurous determination and spirit that led twenty-two-year-old Thomas Kent on his journey to the territory of Montana in 1864.

Creating Montana's Agriculture

Today, the location of Tom and Mary Kent's ranch can be seen from Interstate 90 at the Bridger Creek exit east of Big Timber. A few of the original buildings exist, but they are on private land. The Kent Family Cemetery is nearby on a rise looking over the Yellowstone River Valley. The family continues to have reunions every year if possible at the Bridger Creek home place or at the ranch on Lodge Grass Creek near the Wyoming border. The familiar two-story Kent House, which stood stately for years, is gone now, torn down after a fire. The neighboring town of Reed Point has celebrated the Great Montana Sheep Drive for more than thirty years on Labor Day weekend, remembering the region's success related to the sheep industry, influenced by its leading pioneer Thomas Kent.

In 2019, Tom and Mary Kent were inducted into the membership of the Sons and Daughters of Montana Pioneers. John Spomer recognized that his great-great-grandparents were residing in Montana well before December 31, 1868, the qualifying cut-off date for relatives and ancestors to be eligible.

Winter Splendor takes the chill off of winter as the snow swirls around. Homesteaders were establishing roots and could begin building their homes. Farms and ranches were succeeding. More and more framed houses dotted the landscape. *Courtesy of Gene Roncka, Willow Point Gallery, Ashland, Nebraska.*

6
AT HOME ON THE JUDITH

THE HARDENBROOKS AND THE SKELTONS

Four mountain ranges create a rim around the drainage of the Judith Basin: the Little Belt, Highwood, Judith and Snowy Mountains. The vistas present picturesque scenery with high peaks lingering in the background as backdrops for breathtaking sunrises and sunsets. The Judith River, named by Captain William Clark of the Lewis and Clark Expedition for his sweetheart Julia, flows from its headwaters in the Little Belts though the entire basin to its confluence at the Missouri River.[115] The habitat provides natural protection, quality grass and abundant vegetation. Wildlife thrives in central Montana, home to elk, deer, bear, mountain lion and many other animals large and small.

As stunning as the scenery can be, the climate can be as stubborn. Weather can drastically change in less than an hour. A pleasant day can end with record-breaking temperatures, cold or hot, complicated by gale-force winds. Any type of weather is possible, anytime of the year. Taming the Judith took stamina, heart and soul.

"Open range" ranching was the economic driver east of Montana's continental divide from the mid-1870s until the blizzard of 1886–87. From then on, the cattle and sheep industry saw range management introduced and available land reduced through homesteading. The Homestead Act of 1862 gave an individual the opportunity to own 160 acres providing the land could be put into production and lived on year-round for five years.[116] The railroads connected most of the Montana Territory by 1887, and access to the available claims was made easier for the land seekers.

Settling In and Proving Up

Barb Skelton is a fifth-generation rancher whose pioneer family chose to homestead near the community of Stanford, sixty miles southeast of Great Falls, in the Judith Basin. Skelton continues to operate the Hardenbrook Ranch, which she traces back four generations. Skelton easily reminisces over coffee about the early years growing up in the Stanford area, explaining the biggest part of her family's story is that they stayed. Times were tough, and there were few luxuries or extras—except extra work. They paid the fee to file the claims, broke the sod, farmed, ranched and stayed. They proved up and proved they could do.

Quick to give credit and recognize the strength and fortitude of her dedicated relatives, Skelton related how the family committed to establishing homesteads next to each other. They didn't have a strong background in farming or ranching, but what they lacked in know-how they made up for with their dependable work ethic and the willingness to try and introduce new production methods.

Mr. and Mrs. Fred Miller, Skelton's great-great-great-grandparents, arrived in Great Falls from Winnipeg, Manitoba. Fred's sister Amy Belle Miller joined them too. The Millers' daughter and her husband, Abbie and John Butterfield, would meet them, having traveled from Duluth, Minnesota, where they were involved with a theater company. Formerly of Toronto, Ontario, John Butterfield was a geologist, coming west with intentions of homesteading and mining. From Great Falls, the group traveled south of Stanford to file their claims.

Everyone in the family was involved in the process, both men and women. They had no choice; they were the labor force. They also worked to create a sense of community. Abbie Butterfield was the first midwife in the Stanford area. On Sundays, Abbie prepared dinner and served it on her fine china engraved with silver. She insisted you took time to dress up and enjoy a meal together, a tradition she passed down to their daughter, Bonnie Belle.

Bonnie loved horses, and Barb Skelton is sure she and her daughter got their equine passion from her. After marrying Cap Hardenbrook, Bonnie Belle ranched with and for her parents, while she and Cap pursued their own agriculture interests. They farmed for a short while in the nearby community of Benchland but preferred ranching. The Butterfields and the Hardenbooks are credited with bringing the first polled Herefords to Montana.

A History of Montana Agriculture

Above: Abbie Butterfield, Barb Skelton's great-grandmother, homesteaded with her family next to the Little Belt Mountains south of Stanford and was the first midwife in area. *Courtesy of Barb Skelton*.

Opposite: Harvest time on the Hardenbrook place at Benchland in 1915. Threshing machines separated the grains of wheat from the straw and chaff. The process required teams of horses and men to run all the equipment. *Courtesy of Barb Skelton*.

Creating Montana's Agriculture

Bonnie and Cap's son, John Hardenbrook, continued the ranching and farming for his entire career. A good stockman, John also was an avid and respected hunter from an early age. He married his high school sweetheart, Ruth Crabtree, who joined him in the ranching endeavors. Ruth's father was Harry Crabtree, who worked and was manager for the JB Long Company for over thirty-five years. With more than fifty-four thousand acres, Long had one of the largest sheep businesses in the state. Harry helped bring in the first bunches of black-faced sheep.[117] Skelton shared that her grandfather Crabtree purchased "bucks" from Hank Sieben's Sheep Ranch near Cascade.

As the ranching and farming concerns grew, the families continued to work together. Skelton shared an interesting observation that may have felt unique to her family but wasn't at all that uncommon in the early homesteading and mining days. The pioneer women in her family became widows early in their lives. Their husbands passed away suddenly or from short illness or accidents. As widows, each woman had to make a decision whether to stay or not. It required stepping up and taking the hand that was dealt them. They had to fill the shoes of those who were gone and be able to sustain their operations and maintain their homes.

Skelton reflected that she comes from a long line of strong women: "Their resiliency encourages respect and pride. When things get overwhelming, I have their example to follow, rolling up your sleeves and just digging in." It's the part and parcel of who she's become as she worked side by side with her family

A History of Montana Agriculture

Creating Montana's Agriculture

Above: Four generations of the Hardenbrook family who homesteaded in the Judith Basin. *Pictured left to right*: John Bradford Hardenbrook, Abbie Butterfield, Mrs. Fred Miller and Bonnie Belle Butterfield Hardenbrook. *Courtesy of Barb Skelton.*

Opposite: John Hardenbrook, Barb Skelton's father, was an avid hunter at an early age. Later he would take over the Hardenbrook ranching operation as the third-generation owner. *Courtesy of Barb Skelton.*

Besides overseeing the Hardenbrook Ranch, Barb Skelton with her husband, Paul Gatzemeier, own and run the Intermountain Equestrian Center north of Billings and Horses Spirits Healing Inc. Since 2014, they've featured programing to help veterans recover through the therapy of horses. One visit with Barb reveals her love for agriculture and commitment to preserve the history of her family's stories of homesteading.

THE SIXTH GENERATION

Abbie Skelton, Barb's daughter, represents the sixth generation of ranchers in the Judith Basin through her mother's family, the Hardenbrooks, and through the Skelton family of her late father and Barb's first husband, Stanford William "Bill" Skelton Jr., who passed away on November 25, 2014. Bill's grandfather William Skelton came to Montana from England in the 1850s and to the Judith area in the late 1870s. In 1879, he married Vaitlain Vann, and together they raised eleven children. When William Skelton arrived, the Homestead Act had not been enacted. At that time, land was acquired by the Preemption Act of 1841, which allowed individuals to claim federal land as their personal property by residing, working and improving the land for a minimum of five years.[118] Claimants simply required shaping and placing logs, which were to serve as a foundation for a settler's home.[119] Many others would get additional land by constructing log cabins on other sites. Each "preemption" averaged between forty and sixty acres. William Skelton raised cattle and horses during the free-range period.

One of the more historic events of Montana was the first cattle drive out of the Judith Basin to the railroad in the fall of 1881. The owners of the herd of nearly one thousand cattle included S.S. Hobson, Stadler and Kaufman, George Barrows (whose son wrote the classic book *UBET*), Old Man Belcher, the Skelton brothers and John Duffield. Hatch Tutle was the foreman and Matt Price the trail boss. James Boyer and Harry Keeton were two of the cowboys, and Charles Russell and Frank Plunkett were night herder and horse wrangler.[120]

Charlie Russell, who became an incomparable, beloved western artist, landed his first job in a cattle operation as the night herder, according to Fred Renner, the leading expert on Russell's art. Before the drive got underway at Stanford, some of the cowboys did a little celebrating and tried to ride their horses into one of the saloons, an incident Russell later recorded in his familiar painting *In Without Knocking*.[121]

The Skelton Ranch was a homestead place, settled in 1882 south of Stanford, located at the base of the Little Belt Mountains. Charlie Russell would become one of the family's good friends. Russell spent time in Stanford working for different cattle outfits as a night herder, allowing him to paint during the days. William Skelton is the subject of Charlie's painting *The First Furrow*, which pictures him farming on the bench where the last Judith round-up corrals were located, just south of Stanford. Russell gave the original painting to Skelton.[122] But when the kids got sick with a severe illness there was no way to pay the doctor, William traded it to settle the bill. Three of Skelton's children died from the illness.[123]

Today, Abbie Skelton operates and maintains the Skelton Ranch. Many things have changed since its early beginnings. In looking back, Abbie sums it up well: "A lot of history has passed by this ranch since 1882. I'm proud to continue the legacy."

Morning Mail meant walking down the lane and spending time with Grandpa. Mail was a special event, and in the rural community of Vida it was delivered only three days a week. *Courtesy of Gene Roncka, Willow Point Gallery, Ashland, Nebraska.*

7
HOMESTEADING IN NORTHEASTERN MONTANA

ALFRED AND REGINA WANDERAAS

As the act of homesteading continued in the West, it was soon realized that a quarter of a section of land on the arid Great Plains could never support a family. Changes were made by Congress, and in February 1909, the Federal Enlarged Homestead Act (FEHA) was approved, increasing a claim size to a half section of land—equivalent to 320 acres. By 1912, the proving up period was reduced to three years through the Three-Year Homestead Act.[124] Non-citizens would still have to wait five years to prove up, which encouraged getting citizenship as quickly as possible. To help ensure and increase successful homesteading, the other revision made by the updated 1912 act was requiring the homesteader to live on his land only seven months of the year, which meant he or she could spend the winter months away working to earn money to sustain themselves and put back into developing the land.[125]

The Family Farm

Alfred Wanderaas, coauthor Melody Dobson's grandfather, homesteaded in northeastern Montana in 1910 in the Riverside Community, eleven miles northeast of the small town of Vida in McCone County and eighteen miles south of Poplar. Alfred was seventeen years old when he immigrated to America in 1904 from Steinkjer, Norway, the Trondheim Fjord area where he grew up. He worked in Minnesota until he had the means to take up a homestead claim and chose to file in northeastern Montana.

Homesteading in Montana after the turn of the century was aided by the Great Northern Railroad. This allowed Alfred to travel by rail to the town of Poplar, seventy-five miles west of the Yellowstone River's confluence with the Missouri River. At Poplar, the Redwater River drains into the Missouri River, and a ferrying system was located there to make river crossings.

Wanderaas established his homestead and went to work proving up by breaking the sod and plowing it into fields to plant wheat and oats. On March 12, 1914, Alfred married Regina Christopherson, who had filed in 1911 on the homestead adjoining the west boundary of his claim. Regina was born and raised in Greenwood, Wisconsin, on the family farm of Norwegian immigrants, Peter and Karen Christopherson. Regina homesteaded in northeastern Montana with the support and encouragement of her brothers, Ole, Andrew and John Christopherson, who were also in the processes of homesteading and farming within miles of her claim. During the winter months, Regina worked in Williston, North Dakota, at the Grand Hotel, taking advantage of the time to make preparations, earn money and buy supplies for the return to the claim in the spring. Her letters back home to family and friends in Wisconsin kept them informed of her progress on her claim.

The individual homestead claims of Regina Christopherson and Alfred Wanderaas became one farm when the two were married. Together they moved to the location that Alfred had chosen to put up his homesteading shack and began thoughtfully laying out the buildings and house for the farming operation's homeplace. Alfred went to work planting trees around the west side of the proposed building sites to create a windbreak. A nice grove of plum trees he planted matured nicely. Today, a long row of pine trees catches the eye of visitors as they drive down the lane into the yard. Alfred and Regina raised their family of six children—Arthur, Harold, Nellie, Helen, Bernard and Leonard—on the homeplace, and the farm is still in operation and has been for over a century.

Raising and supporting a family and helping them launch into their livelihoods brought true joy to Regina and Alfred. Their children would reminisce with fond memories of growing up on the farm and sharing in the responsibilities of the chores, the home and the farming until it was time to further their education, embark on careers or serve in the military. Letters kept them connected, and the trips back home were always celebrated reunions. Hardships were a reality to endure and something they got through together.

A History of Montana Agriculture

Alfred Wanderaas and Regina Christopherson were married on March 12, 1914. Each had homesteaded their own claim south of Poplar in the Riverside community of northeastern Montana. They joined their homesteads into one farm after they were married. The farm is still being run by family members today. *Courtesy of the Wanderaas Family Photo Collection.*

In the early years, when their children were younger, the couple lost their ten-year-old son, Harold, to a childhood disease. The love from family and support of their neighbors and friends helped sustain them through the good and bad times. Being there for one another was an unspoken understanding of life on the prairie.

In 1946, Alfred's son Bernard joined him in the farming operations. Bernard and his wife, Ruth, built their house on the homeplace and raised their family of seven: David, Camille, Bonnie, Steven, Melody, Bernadine and Sylvia. Upon Bernard and Ruth's retiring, son Steve invested in farming and ranching, which included the Wanderaas family farm, and he lives on the homeplace with his family. Most of the original buildings, including Alfred and Regina's house, are being used by the third and fourth generations as they continue to farm.

Ole Christopherson, Regina's older brother, was one of the first homesteaders to arrive in the Riverside area and would eventually help bring all his siblings west to homestead in Montana and North Dakota. Never marrying, he assisted friends and neighbors in succeeding on their claims or making the transition when it didn't work out. Being good at carpentry, Ole offered his constructions skills and built several homes, plus outbuildings that still exist on original homesteads. Pieces of cabinetry and furniture he crafted remain with the family.

Building Community

Alfred and Regina Wanderaas, along with her brothers, sisters and their families, worked together through the tough seasons of the Great Depression, droughts and dust bowls. Alongside their neighbors, they built a community—schools, churches and developed roads. Regina donated a portion of her land for what would become the location of Southview School. The country school was in operation until its closing in 2001. A celebratory reunion was held to commemorate all the families of students who attended the school from the time it was established in 1919.

Creating an environment for social gatherings and friendship helped the homesteaders deal with heartaches, hardships, learning the English language, losses and loneliness that accompanied the trials of homesteading. Very few would return to their homelands or hometowns again, and it was through letters, newspapers and occasional visits that kept the ties to loved ones back in Wisconsin and Norway.

Author Melody Dobson recalls accompanying Grandma Regina as a little girl to visit Regina's neighbor, Mrs. Olson, as the ladies shared coffee and talked about a letter recently received from the "old country." Lifetime friendships were built into the next generations over pots of coffee, visiting, card playing, picnics and working shoulder-to-shoulder together to sustain their farms and to survive.

Homesteading kitty-corner from Regina's claim, to the southeast was her friend Alida Swenson, also from Greenwood, Wisconsin. The two women continued their friendship on the eastern prairies of Montana. Many of those staking claims right around them were also from Greenwood. Those making the move were the daughters and sons of families who had emigrated years earlier from Norway to begin their lives in Wisconsin, which reminded them of the woodlands and climate of their native country.

In 1911, Alida became a teacher at one of the first schools established in the area after borrowing books from Poplar to study for the Montana teacher's exam. Hilger School, located near Redwater River, was founded in 1908 after the homesteaders petitioned the commissioners of Richland County three times. Before the schoolhouse was built, the first classes were held in Nels Tolan's log cabin.[126]

Hilger School was located over eight miles from Alida Swenson's claim and sometimes called the Riverside School since the Riverside post office was started one mile away. During the winter of 1911–12, Alida stayed at the home of Mary and Charles Hilger, where the temperature was recorded to have reached fifty-two degrees below zero.[127]

Alida married Pricy Vine, who homesteaded a claim just south of hers. Known as Price and a good friend of Regina's older brother Ole, he arrived in the area at the same time. Regina Christopherson and Alida Swenson each successfully filed claims of 320 acres, allowable through the updated provision of the original Homestead Act of 1862, the Federal Enlarged Homestead Act (FEHA). Both ladies proved good on their claims.

Regina and Alfred Wanderaas and Alida and Price Vine were close friends and neighbors. They supported one another, gathering often for work and play. In a letter Regina wrote to her son Leonard on December 14, 1959, she noted that Alida and Price had come to visit her and Alfred to celebrate her birthday. The letter also shared that she received a new granddaughter as a gift on her birthday. Coauthor Melody Dobson celebrates the same birthdate as her grandmother Regina.

Right: Regina Wanderaas sits by the heating stove in her and husband Alfred's new home. The house was warm and inviting, as the couple regularly hosted family and friends. *Courtesy of the Wanderaas Family Photo Collection.*

Below: Alida Swenson on her horse, stopping next to her homestead shack in northeastern Montana. Alida's homestead claim was kitty-corner southeast to Regina Christopherson's claim. The two women were good friends and had lived in the same community in Wisconsin before moving west to Montana. *Courtesy of the Wanderaas Family Photo Collection.*

Author Melody Dobson recalls celebrating one of her early childhood birthdays with Regina and Alida. "It was a coffee party to celebrate our birthdays, and Grandma Regina took me with her to Alida and Price's home. Waiting for me on the table was a carefully wrapped package. Inside were crayons—my first box of brand new—sixty-four colored Crayola Crayons from Alida. It's a gift I'll never forget."

The Wanderaas and Vine children and grandchildren continued friendships over the years and miles. Bernard Wanderaas made the decision to farm and began helping his father, Alfred. Price's son John made the decision to farm as well and began helping in the Vine family's farm operation. As young men, both Bernard and John saw and helped their parents through tough years of farming, as John recalled in an interview in 2001:

The early 1930s were dry and tough. Later on in 1938, a lot of the area had fairly good crops. My parents took out hail insurance that year, as did the Wanderaases and Loendorfs. As it turned out, Loendorf's crop was hailed out and they collected the insurance, but the grasshoppers ate up Wanderaas's and our crops. We paid the insurance and still didn't have a wheat crop.[128]

The families of Alfred and Regina Wanderaas and Price and Alida Vine gather together for a photo, circa 1930s. The two families were neighbors and close friends throughout the years, helping each other build their homesteads into successful farming operations still being farmed by their respective families today. *Courtesy of the Wanderaas Family Photo Collection.*

Creating Montana's Agriculture

With no crops or extra funds, the Vine and Wanderaas parents were not able to send their sons John and Bernard to Wolf Point to attend their first year of high school. Instead, the young men were enrolled into a correspondence course with the University of Nebraska–Lincoln, administered through the McCone County Superintendent of Schools in Circle. Classes were held at Southview School, the elementary country school close to their family farms. The Southview School teacher and the county superintendent of schools oversaw their studies.[129] John and Bernard would later be able to attend and finish high school at Wolf Point High School. Both graduated in the top ten of their class in 1942; the class motto was "No Victory Without Labor."[130]

Wolf Point was twenty-five miles north from the Wanderaas and Vine family farms on the north side of the Missouri River, in Roosevelt County. The families and their neighbors did a majority of their business and trade in Wolf Point, which continued to grow and sustain itself as a strong agriculture center. Originally a trading post, Wolf Point grew as a new site on Jim Hill's Great Northern Railroad. Hill was pushing to get the surrounding lands open for homesteading, and by 1909, the new town had begun to prepare for the mass of homesteaders soon to arrive. Author and local historian Marvin Presser wrote in his book *Wolf Point: A City of Destiny*, "There were land rushes before, but there was never anything like the rush to settle Montana nor anything like the collapse that came just a few years later."[131] According to Presser's narrative, the old-timers gave Wolf Point the nickname "Norway" because so many people flooded in from the Scandinavian country.

With railroads, roads, new highways soon to be built and an impressive bridge over the Missouri River commissioned on July 1, 1930, Wolf Point was a key location and convenient place to deliver agriculture products and pick up supplies. Marvin Presser cited notes in his book about each year's growing season, the harvest and notable developments of agriculture that took place:

> *1920—Farmers concerned with the loss of moisture from their fields and the many thistles, were highly interested in the new tillage machine being demonstrated by Ben Balerud, manager of Champlin Implement. The new machine was called a "rod weeder." It seemed to kill weeds and also conserve moisture.*[132]
>
> *1921—In August, harvest began in earnest. Crops were running from a low of nineteen bushels per acre to a high thirty-six bushels.*[133]

Wolf Point was a major agriculture hub where families did a majority of their trade. Farmers hauled grain to the elevators on Front Street after the Lewis and Clark Bridge was completed on July 1, 1930, over the Missouri River. Alfred and Regina's children and grandchildren attended high school in Wolf Point. *Courtesy of the Wanderaas Family Photo Collection.*

> *1929—In late June, the Wolf Point Commercial Club made sure the state press was aware that local elevators had shipped 1,442,256 bushels of the 1928 wheat crop—a new record.*[134]

Before settling in to work full time in their family farm operations, both Bernard Wanderaas and John Vine enlisted in the military during World War II. Bernard deployed to the navy, and John served in the army. During the service years, young men and women with agriculture backgrounds learned skills that served them well once they returned to work on their respective farms. Mechanical technology was advancing rapidly as the United States put forth its best ingenuity to fight the war in Europe and in the Pacific theater. It also taught the service men and women resourcefulness and to learn how to make do and survive when supplies were limited and rationed. Both Bernard and John demonstrated the patriotism and pride of citizenship that their experiences instilled in them, and they passed the sentiments on to their children and grandchildren.

Returning after World War II, Bernard Wanderaas and John Vine each became involved in farming their parents' farms and remained good friends and neighbors. Bernard and John raised their families, and the next generation began to take interest in farming too. Bernard and Ruth Wanderaas's son Steve began working in agriculture right after high school. Steve also enjoyed ranching and continued developing the livestock component of the farm that Bernard had started. John and Marilyn Vine's son Leonard chose farming as a full-time career but first attended Montana State University studying computer science before returning to begin working on the family's farm. Fourth-generation family members are now beginning to become involved in the farming operations as well.

In 2011, a great-granddaughter of Regina and Alfred Wanderaas, Brianna Bohmbach, married Kyle Vine, the great-grandson of Alida and Price Vine. Brianna and Kyle are involved in agri-business and make their home in Vida. Their children, Landry and Tatum, are the fifth generation being raised near the original family farms and homesteads.

When Alida Vine's grandson Bob assisted her in preparing to write her story, "320 or Bust," little did he know that he would be helping leave a "lasting legacy."[135] Her historical memoir provides insights and descriptions into the early homesteading years when pictures were not available to reveal the rest of the story. A copy of the manuscript is in the Montana section of the George McCone County Library in Circle.

A New Era

The family farm saw the transfer of field work and labor being done by teams of horses to the invention of steam engines and affordable tractors. Adapting and adjusting for these changes was constant. The new workhorse on Alfred and Regina's farm was their Twin City tractor manufactured by Minneapolis Steel and Machinery Company. Named for the famous twin cities of Minneapolis and St Paul, the company's original logo resembles the Minnesota Twins major league baseball team's logo. The Twin City tractor was accessible for the small farmer, versatile and capable of breaking the prairie ground.[136]

> *Alfred's grandson Steve Wanderaas visited with author Melody Dobson, his sister, and shared how Grandpa Alfred and Grandma Regina survived during World War II: "They had purchased new updated equipment before the war started. That was really one of the things that saved them. Once*

A History of Montana Agriculture

This page, top: Tractors were the new workhorses. A tractor could do in days what the draft horses would do in a week. Homesteaders slowly mechanized their farming operations, but horses were still used and were a valuable asset. *Courtesy of the Wanderaas Family Photo Collection.*

This page, bottom: Harvest time on the Wanderaas farm. Here the threshing machine is powered by the tractor, while the wagons hauling the bundles of wheat are pulled by horses. *Courtesy of the Wanderaas Family Photo Collection.*

Opposite: Henry Ford's Model T trucks from the mid-1920s were a welcomed addition to production agriculture and made hauling grain easier and less time consuming. *Courtesy of the Wanderaas Family Photo Collection.*

Creating Montana's Agriculture

the war started you couldn't purchase anything or get parts or tires. The war happened right when lots of folks were ready to update and they just couldn't afford to. Grandpa and Grandma had a car, and they had just bought a 1941 Dodge pickup that did a lot of work and that was their truck. The new equipment was reliable and really helped with the hard work around the farm while Dad was away serving in the navy. When the war ended, one of the first things they had to do was go and buy new tires for the tractor."

When Bernard Wanderaas returned after the war, he began his livelihood as a farmer immediately, in 1946. The postwar times brought new research, technology and education for agriculture, and Bernard was asked to assist and lead a vocational program for veterans at Wolf Point High School. During that time, he met his future wife, Ruth Blair, the home economics teacher. After marrying, the couple became a part of the farming community while working and raising their family.

Being innovative and adaptable to change was helpful in the farm's progression. Regina Wanderaas reflected in one of her letters in the mid-1950s that Bernard had just built hog pens and was going to introduce pigs.[137] The advances in animal husbandry, soil conservation, machinery, and commodities allowed for the opportunity to diversify and implement new ideas. Even though the farm consisted mainly of the same acres since 1910, the landscape continued to change with new additions and improvements.

A History of Montana Agriculture

Alfred and Regina Wanderaas's son Bernard prepares to deploy for the navy during World War II. Many of the immigrant pioneers saw their sons and daughters leave home to fight in the war and worried about the tensions in their native countries with concerns for family members still living there. *Courtesy of the Wanderaas Family Photo Collection.*

The Wanderaas farm has stayed in the family for four generations and for over a century. Alfred and Regina's son Bernard returned from World War II to farm with his parents. Upon retiring, Bernard turned the mantle over to his son Steve and his family. Paul is involved with the day-to-day farm and ranch operations and has chosen farming as his career while Mark and Travis are involved and work in the farm equipment agribusiness, offering their knowledge, support and encouragement throughout the year. *Front row, left to right*: Steve Wanderaas and Bernard Wanderaas; *back row, left to right*: Travis Peterson, Paul Wanderaas and Mark Peterson. *Courtesy of Cathy Wanderaas.*

Transitioning the farm from one generation to the next is called succession management and is an important part of sustaining the family farm. As the third-generation farmer and rancher, Steve began working on the family's dryland farm during the challenging 1980s. He admits that farming is never without its economic challenges of up-and-down commodity markets and unpredictable weather, but the opportunity has been rewarding.

Being able to expand, diversify and manage those changes is a part of the daily operation. It's important to listen to the information and education being provided on new products. Spending time reading, asking questions and watching the markets all play into the agronomy of making decisions whether or not to try a new commodity and implement something new. The introduction of pulse crops, mainly peas, has been good for the farm. Even though the markets have fluctuated, it has been good for enhancing the soil.

In encouraging the next generation, Steve reminds them to "be open-minded," and if it's something you want to do, "pursue your passion." Steve and his wife, Cathy, live on the original Wanderaas farm, where their son Paul joins them in the day-to-day operations and has started his own farming career as the fourth generation.

Demonstrating Responsible Stewardship

Steve Wanderaas takes an active role in being a good steward in managing the land that he farms. Soil conservation is important in protecting the ground and water resources used to provide the farm's income, especially in dryland farming. Steve has been an active participant in his local county Conservation District for many years. In fall of 2019, Steve was awarded the Montana Association of Conservation Districts Supervisor of the Year award for his leadership of McCone County Conservation District.

Serving on the McCone County Conservation District, Wanderaas knows the need to monitor the threat of invasive species. He works with the Montana Department of Natural Resources & Conservation's Montana Invasive Species Council (MISC). He's been involved since 2016 and has served as MISC's co-vice-chair, helping in the efforts to educate other Montanans on the direct impact invasive species has on agriculture, the ecology and economy of the state.

Along with managing the farm, Wanderaas finds time to participate in meetings, often driving hundreds of miles to attend. Current threats to the states ecosystem include the aquatic invasive species Eurasian watermilfoil and New Zealand mud snails, which have found their way into Montana's water bodies.[138] If left unchecked, severe damage affects water resources used for irrigation and recreation. Steve shared,

> If these mussels get into a lake, they will wipe out the lake. Once the food chain is gone, fishing is gone and recreation is gone as well. A beautiful beach once enjoyed will be destroyed. With no more boaters or people enjoying the beaches, the once relied upon economy will also be gone. That will affect the culture and the dynamics. It changes history.

Remarque of *Back Roads. Courtesy of Gene Roncka, Willow Point Gallery, Ashland, Nebraska.*

A statewide campaign is in effect to provide public awareness and alerts to make sure all watercraft users clean, drain and dry their boats, canoes, kayaks or any type of equipment after each time it has been used in the water. Checkpoints have been set up to inspect all boats going in and coming out of the water.

Besides aquatic invasive species, there are new invasive species that will affect grasses, trees and wildlife. The emerald ash borer threatens ash trees.[139] Wanderaas cautions, "Imagine a city park without the beautiful ash trees." Being part of a farm and ranch operation helps Steve realize how susceptible resources are and how something like an invasive species will hamper the ability to produce and generate income. Wanderaas concluded that awareness is key and staying alert will help keep all Montanans including farmers and ranchers, stay ahead of the impacts on the state of Montana.

8
THE HOMESTEADING BOOM

The homesteading era brought people in search of a new beginning and others wanting to take advantage of the chance to get rich quick just like the advertisements bombarding the newspapers implied. The Federal Enlarged Homestead Act of 1909 set off "an explosion of promotional campaigns aimed at luring settlers."[140] The railroads were the biggest advertisers, offering cheap rail fare in America and Europe. Large tracts of land were purchased by the railroad companies to ensure the railways could be built, and now they offered the acres back at next-to-nothing rates. The railroad owners and investors were anxious to see the land developed and growing crops so they could haul the grain with their boxcars that stood empty.

The biggest promoter was Jim Hill, the builder and personal financial backer of the Great Northern Railroad. Hill was focused on recouping his money and making the railroad pay for itself. At the turn of the century, Hill encouraged and created experimental farms to promote agriculture. Given enough time, the small test plots on the farms produced beautiful sheaves of wheat and test samples of grain to be exhibited at fairs and events in the eastern states and in the Midwest, as displays of what was being grown in the West. Hill used propaganda to promote and fill the "last underdeveloped portion of the United States—the Great Plains of Montana, and of North and South Dakota."[141] Billboards lined the railroad's right of way depicting farmers plowing money out of the earth and advertising that millions of acres of fertile land could be "had for the taking" in northern Montana.[142]

HOMESTEADING: LESSONS LEARNED

Milburn Lincoln Wilson, known as M.L. Wilson, served as undersecretary of the U.S. Department of Agriculture from 1937 to 1940 for both President Franklin D. Roosevelt and President Harry Truman during the years of the New Deal. During Wilson's youth and through college, he desired to homestead; came to Fallon, Montana; and believed he could successfully file a 320-acre claim in 1909. Wilson gained intuitive insights that he used later as the undersecretary to draft the major farm policies for the Great Plains.

Homesteading was a speculative move, and Wilson spent a year calculating his decision, planning, preparing and finally selling his 240-acre livestock operation in eastern Nebraska. He graduated in 1906 from the School of Agriculture at Iowa State in Ames, Iowa, where he majored in agronomy, the predecessor of modern farm management.[143] After traveling to Miles City to file his claim with the Federal Land Office and waiting in line for several days, his view of homesteading was altered forever. Wilson wondered if homesteaders had a place in rational agricultural development.

> *The hotels were overflowing, and Wilson spent his nights on a makeshift cot in a large room "jammed with seekers." Teachers, laborers, miners, lawyers, even gamblers and dance-hall girls rushed to get some of the available land. The one belief they all shared was that they could make fast money. Having read glowing stories in hometown newspapers, railroad fliers, and even the* Saturday Evening Post, *they had stormed into Miles City for their share of the anticipated windfall.*[144]

With determination and confidence in his professional training in agriculture, agronomy and management, Wilson took on the challenge of farming the plains. He now had his own claim and a land partnership he arranged while still living in Nebraska. But the weather and nature were against him. The ventures failed. What Wilson soon realized from his experiences is that there was a need for education in farm management and agricultural mechanics, which the homesteaders lacked and needed.

Wilson inquired about working with the College of Agricultural at Montana State College and in 1911 became the superintendent for the college's twenty experimental farms in eastern Montana. His new job also included the responsibility of serving as the extension agent for Custer County. Wilson knew, "The Plains were different and required new departures in farm mechanization, management and economy."[145] In his positions, Wilson worked hard to make a difference and was successful in seeing new methods introduced. His keen understanding served the agrarians of the plains well, as he devoted his lifetime to this endeavor as an educator and as a statesman.

The End of an Era

For those who could survive the elements and economics of proving up, there was success when rains came at the right times. Hope sprang eternal on the prairie with amber waves of wheat. Crops produced enough to feed families, animals and the markets. Optimism ran high, and investments were encouraged with credit easily available. By 1918, Montana's population tripled in less than twenty years to 769,590.[146]

Eventually, the push to plow the prairie took its toll. The rain stopped falling over parts of northern Montana in the spring of 1917, and the following year, drought spread throughout all of eastern and central Montana. The crops were decimated. Grasshoppers cleaned up anything that was left. Wheat prices bottomed out in 1919 after World War I. The significant growth that occurred during the boom made this loss "so pronounced and so astounding that when it all came crashing down, the enormity of the tragedy was incomprehensible."[147] These catastrophic events marked the beginning of the end of the last migration attempted to develop the prairies of Montana.

In 1909, the yearly outlook appeared so promising. The growth that occurred for the next ten years could not sustain itself or outsmart the weather. To worsen the situation, it became apparent that a good portion of the land designated to be claimed was actually not even suitable for farming. "From 1910 to 1922, homesteaders located on 42 per cent of the land in the state, yet only 20 per cent of this land was actually fit for anything but grazing. Such was the effectiveness of Jim Hill's bold advertising."[148]

Creating Montana's Agriculture

Geese in Flight pictures a common scene on the northeastern Montana plains, which provides countless natural wildlife and bird habitats. Perhaps the homesteaders took solace in and identified with the migrating patterns of their feathery friends. *Courtesy of Gene Roncka, Willow Point Gallery, Ashland, Nebraska.*

The Homestead Act signed by President Abraham Lincoln in 1862 gave away over 10 percent of United States land for free in exchange for a person's commitment to a portion of land identified as a claim and proving he or she could work the parcel for three to five years. Montana led the way in proving up 151,600 homesteads, the most in the nation, while North Dakota reported 118,472, Colorado 107,618 and Nebraska 104,260.[149]

Part 2
GROWING MONTANA'S AGRICULTURE: INVENTIONS AND COMMODITIES

9

THE MIRACLE INVENTION

When authors Jody Lamp and Melody Dobson met Roland Temme, founder and owner of TMCO Manufacturing in Lincoln, Nebraska, he disappeared for a few minutes to find an item of intrinsic sentimental value to share. He knew the two came from homesteading agriculture backgrounds and wanted to see if they could guess the item's identity. The small object was mounted on a round base that fit in the palm of his hand. They didn't know, so Temme explained it was a "knotter" from a binder.

The binder was invented by Charles Withington in 1872 and was rated as one of the greatest inventions to help the farmer with this harvest.[150] The binder could be pulled by a horse or tractor, and it helped cut the crop mechanically by forcing the stalks of grain into a sickle or sharp moving metal bar that cut the stalks. They then fell onto a moving canvas serving as a conveyor belt to carry the grain to an arm, where the stalks would be tied into a bundle. Then, when five to seven bundles had been created, all of them together would be dropped to the ground. The bundles would then be shocked, or stood up against one another, so they could easily be grabbed by a pitchfork and thrown into a bundle wagon. The grain bundles were then fed into the threshing machine, which separated the grain from the stalk and the chaff, completing the harvesting cycle. The binder and threshing machine would later be combined into one machine, creating the combine we see harvesting grains in the fields today.

A History of Montana Agriculture

The binder was an essential piece of equipment on the farm. Here Alfred Wanderaas uses a four-horse team to pull and operate the binder through fields of wheat on his homestead in northeastern Montana. *Courtesy of the Wanderaas Family Photo Collection.*

Combines replaced the threshing machine, and the newer models were pulled by tractors instead of horses. The new farm trucks hauled more grain faster and farther. This scene depicts a light-hearted harvest moment enjoying the new machinery. *Courtesy of the Wanderaas Family Photo Collection.*

Nicholas Schuman's combine was the first to arrive in Montana. Manufactured by Oliver, the Nicholas & Shephard model was pulled with Schuman's team of horses to cut and thresh his crop at Wheat Basin. This would be the first time harvest would be done with only one piece of machinery. *Courtesy of the Ted Schuman family, descendants of Nicholas Schuman.*

The knotter, as Roland Temme explained, was the little piece of equipment that would hook and tie the knot once the twine was around the bundle of grain in the binder. Temme said it was known as a "miracle invention" and was the technological breakthrough of its day. Simply tying the knot by machine saved so much time. Early bundles were bound with wire, which tended to leave bits of metal in the hay and grain. The knotter was invented by John Appleby, who demonstrated it first in 1867—it was later adapted.[151]

> *As a young girl, author Melody Dobson drove the tractor for the binder owned by a relative who still wanted to cut a field of oats the "old-fashioned way." Later, she and her siblings shocked the bundle piles.*

The combine did exactly that; it combined cutting the wheat stalks and feeding them into the attached threshing component and collecting the grain kernels into a hopper or holding tank to unload at one time. The combine was originally pulled by horses and soon by tractor. Nicholas Schuman brought one of the first combines into Montana. Schuman farmed at Wheat Basin. His Oliver combine was the first to cut wheat in the area.

The 1920s and '30s proved to be difficult years in Montana. Droughts plagued the plains and the overgrazed ranges, and the depleted cultivated lands brought the dusty thirties. But during this time, trying as it was, the challenges forced the beginning of change, and those in the industry knew the "long road to agricultural prosperity would demand major adjustments."[152]

10

THE MECHANICAL REVOLUTION

The industry of agriculture also changed. The mechanical revolution for agriculture was introduced after the turn of the last century, and enormous advances were ahead for the agrarian as the source for horsepower transformed into the invention of the mechanical engine. The workhorse was replaced by tractors; it was easier to supply gas and oil for a tractor than hay and water for animals. Farmers could shut off the engine and, without further ado, go into supper at the end of the day.[153] Small internally geared gasoline tractors went to work on the farm, and by 1930, nineteen thousand were at work on fourteen thousand Montana farms.[154] The farmer could realize more work in one day and utilize more acreage for production without having to use land for feeding and pasturing workhorses. Trucks were available for general use on the farm, and most farmers used Model T trucks for hauling by 1925.

When asked what he deemed the most significant mechanical development other than replacing the horse, Rodney Broderson, a well-known and successful mechanic in the Yellowstone Valley, replied without hesitation, "Henry Ford's invention of the mass production of parts." Most farmers were able to get Model Ts, but the availability of parts to fix the breakdowns meant being in or out of business.

Technology advancements assisted the farmer and rancher in every way, easing their workloads by speeding up laborious processes and helping with the everyday business management of the operation, including interesting

A History of Montana Agriculture

In 1925, this Model truck was the first to haul sugar beets over a beet dump in the Great Western Sugar district on Huntley Project. *Courtesy of the Huntley Project Museum of Irrigated Agriculture Photo Archives.*

innovations that helped increase production. More agribusinesses began to emerge in the small rural towns and communities that could focus on supplying and selling the new parts, tires and equipment needed to keep the machinery going. Implement dealers and their mechanic shops were regular stops on any trip to town.

Production agriculture continued to expand in the 1970s as the growers increased their operations by converting ranch land into fields and with the purchase of more property. More powerful equipment was needed to work and till the land within the five-to-six-month growing season of the High Plains. Equipment was invented and engineered to get the job done. Ron Harmon, owner of Big Equipment in Havre, left a legacy in these advancements. In 1975, Harmon began manufacturing the biggest tractors in the world. The Big Bud brand of tractors filled a void in production agriculture. Through years of ups and downs in the manufacturing and distribution world, Harmon vested himself in developing the Big Bud tractors. When manufacturing of the big tractors ceased, he turned his focus on servicing, rebuilding and refurbishing the different models still available and in use.

Right: Cecil and Ruth Zody homesteaded in the Bloomfield area northwest of Glendive. They operated their farm from 1914–1945, when they moved into Glendive. Cecil became a partner in a thriving business the community referred to as Birdsall & Zody. The enterprise focused on farm equipment and tires. *Courtesy of the Terry Dobson Photo Collection.*

Below: Ed Lenhardt is harvesting sugar beets with his new six-row harvester west of Billings by the town site of Hesper, once an active agriculture community. Lenhardt praised the upgrade from the one-row harvester, appreciating how the innovation made harvesting go much faster. *Courtesy of the Ed Lenhardt family.*

A History of Montana Agriculture

Wilbur Hensler and Bud Nelson, also from Havre, worked on innovations for the big tractors that resulted in a four-wheel drive workhorse for the prairie that could boast an engine of over 700 horsepower. It was Harmon's intuitiveness and mechanical know-how along with company engineers that turned the men's applications into the branded series of great equipment known as Big Bud. The largest tractor built by the company was the Big Bud 747, with the capacity of a 760-horsepower engine. The tractor could easily plow ninety feet at a time.

> *The creation of The Big Bud V16–747 (named after the Boeing 747 jumbo jet) came from a challenge inspired at an agribusiness trade show in California known as the Tulare Farm Show, and Ron Harmon was there marketing the Big Bud tractors. In a personal interview with Ron Harmon at his Big Equipment office in Havre, he recalled how the Rossi brothers approached him to see if he could build a tractor big enough that could respond to their needs in 1977. Harmon answered the challenge with the 747 model. The Rossi brothers farmed near Bakersfield, California, with equipment from Harmon and Caterpillar D-9 tractors. What they would accomplish in one day with their equipment, the new Big Bud 747 knocked out in one hour.*
>
> *Ron explained how the big tractor eventually made its way back to Montana when brothers Randy and Robert Williams from the Big Sandy area decided to purchase the Big Bud 747 when they learned the tractor was available. They restored the tractor to "farm ready." Ron laughed when sharing he never realized the popularity and interest the tractor had. People came from all over to see it, even Europe. Now there is a website to buy miniature models. The Big Bud V16-747 has a one-thousand-gallon fuel tank, weighs over sixty-five tons when it is full of fuel. It's twenty-seven feet long, twenty feet wide and fourteen feet high. It has the ability to pull an eighty-to-ninety-foot cultivator at seven miles per hour.*
>
> *At the close of the interview, Harmon shared a copy of an article from the* Hastings Tribune *in Nebraska dated July 10, 1981, describing the Nebraska Tractor Test Laboratory session on testing the Big Bud 525 HP Model. The Tractor Testing Lab is part of University of Nebraska–Lincoln's Institute and Natural Agricultural Research Division, located on the university's East Campus. The big tractor was over eight towers standing shoulder high and sixteen feet wide, one foot wider than the Tractor Test Lab's testing track. To solve the dilemma, the university's lab made the arrangements to have the Big Bud tested using Lincoln's old air base. The*

Ron Harmon led the way in developing and marketing the Big Bud Tractor line from the mid-1970s through the 1980s. The Big Buds were manufactured in Havre and shipped nationally and internationally. The Big Bud V16-747 is the largest agriculture tractor in the world. *Printed with permission from Ron Harmon. Artist, Don Greytak.*

> *Big Bud 525 HB is the tractor with the highest drawbar horsepower of any model tested in the history of the University's Tractor Testing Laboratory. The article noted the tractor would be on display at the UNL's field lab in Mead on July 23, 1981.*
>
> *When asked at the close of the interview what accomplishment he was most proud of in developing and manufacturing the Big Bud line of tractors, Harmon thought for a moment and replied, "I think it's keeping the ag business, in terms of tractors, serviceable, re-buildable, and infinitely updateable for the farmer."*

Leland Hintz, known as Lee, purchased a used Big Bud tractor in 1978, valuing the versatile innovations the new tractor offered. Hintz would share with his son Brent Hintz how it was necessary to add equipment to the farm that they could work on themselves. The enormous size of the new tractor was not intimidating to Lee as long as he could access and mechanically

Lee Hintz stands in front of his HN 320 Big Bud tractor. Lee was the second generation to operate their family farm south of Wolf Point on Highway 13 near Vida. Hintz purchased the used Big Bud tractor in 1978. The tractor offered four-wheel drive and lots of power plus the versatility to allow Hintz to make repairs and work on the machine himself. The Hintz farm continues in the family today as daughter and son-in-law Lee Ann and Richard Rush farm the original homestead grounds. *Courtesy of Lee Ann Rush.*

make major repairs himself. Brent recalled the details of the tractor, a two-year-old HN 320 model built up to a 360-horsepower engine. The rebuild had been done in Havre, where Big Bud manufactured its tractors. The younger Hintz remarked, "Dad liked that we could get good parts anywhere and if we had to we could use truck parts." Lee Hintz's ability as a mechanic was a great asset to their farm and demonstrated how keeping the operating machinery in good repair and serviced resulted in great savings of money and time. In the years before Lee purchased the Big Bud tractor, he was one of the first farmers to retrofit and combine other tractor parts and kits to build his own four-wheel tractors.

The history and evolution of farming techniques and equipment is displayed and interpreted at the Montana Agriculture Center and the Museum of the Great Northern Plains in Fort Benton. The museum features the generation of settlers who came and opened the plains up for production. Since 1989, with the help and vision of museum curator Jack Lepley, the interpretive center's engaging displays have shared the amazing stories of the state's first agrarians. The museum sits in the center of history near the Missouri River at Fort Benton; the town was the inland port that opened up the West during the nation's early expansion and immigration years.

11
LEARNING TO GROW IN DIVERSE CONDITIONS

Many of those who came to Montana to homestead had some sort of agriculture background, but most were not prepared for the arid dry lands and climate of the prairie. In addition, understanding the different dryland farming techniques and equipment presented challenges while irrigation was being developed and introduced by trial and error.

The U.S. Bureau of Reclamation introduced the Reclamation Act of 1902, which paved the way to create irrigation systems and construct holding dams for the purpose of bringing water to the arid lands, thus bringing the land into agricultural production through the method of irrigation. The first project completed by the bureau was the Path Finder Dam in Wyoming holding water for the North Platte Irrigation systems of eastern Wyoming and western Nebraska. Montana would receive the funds to have six major irrigation projects.

Construction for the Huntley Irrigation Project Ditch began in the spring of 1905, and on June 26, 1907, the ditch was completed and the gates opened, ready to bring water to thirty-five thousand acres. The cost was $900,000.[155] With water available and land approved for homesteading by a survey done in 1903, the area was ready for new landowners to seek their claims.

Neal Gunnels, executive director for the Huntley Project Museum of Irrigated Agriculture, explained how the new lands were opened in the district for settlement. Those who wanted to file had to enter into a drawing that would take place on the day the ditch opened and all those intending to enter had to have their registrations in the day before and pay the twenty-five-cent fee. For the 1,000 names that would be drawn,

Pompeys Pillar in 1914. Captain William Clark ascended the sandstone bluff overlooking the fertile Yellowstone Valley 108 years earlier. Wildlife could be seen in all directions as far as the eye could see. *Courtesy of the Huntley Project Museum of Irrigated Agriculture Photo Archives.*

The irrigation ditches and canals were built to water the crops on the Huntley Project Irrigation District, and they require regular maintenance and repair. The Ruth Dredger Manufacturing Corporation supplied the earlier models of dredgers used by the district. An original Ruth Dredger is displayed in front of the Huntley Irrigation Museum outside of Huntley. *Courtesy of the Huntley Project Museum of Irrigated Agriculture Photo Archives.*

5,491 people registered. James Garfield, the secretary of interior, started the drawing at 9:00 am on June 26, 1907, at the land office. After the drawing, a special train would take over 1,000 people to Huntley Project for the grand opening of raising the gates on the new ditch. Starting on July 22, 1907, the first 50 names drawn were allowed to file, followed by 50 names each day after until the list was gone through. Many of those who entered the drawing never intended to win but only wanted to see if their name would be drawn. Of the first 1,000 names drawn to receive a land claim, only 76 people filed.[156]

Homesteads slowly began to appear on Huntley Project, and the water wouldn't actually be available to the growers until spring of 1908. Getting a foothold was a process of learning and working together to get the land cleared of sagebrush and tilled so they could begin planting. Since the 1870s, agriculture experimental stations and farms were in use in Montana and yielded useful and proven information from the trial-and-error experiments conducted and recorded each year. Data was readily available, and the new knowledge helped producers learn how to grow a successful crop or the best and most productive, profitable usages for the acres. Positive results were made and replicated continually.

Growing Montana's Agriculture

In order to learn how and what to grow specifically on the Huntley Project, an arrangement was made with the Montana Agricultural Experiment Station to establish a demonstration farm on the Osborn town site. An appeal was made to I.D. O'Donnell to come and direct the efforts since he had worked in the promotion of agriculture practices and had applied them to his farm operations west of Billings.

> *I know it would profit the settlers very much to have this demonstration farm operated by you, in order that they might have the benefit of your experience and example, and also opportunity to consult with you about the crops growing under your management and their own within the project.*[157]
> —*H.N. Savage, supervising engineer*

I.D. O'Donnell worked hard as a farmer and leader in the development of agriculture in the Yellowstone Valley and nation. He vested himself in the principles of new agriculture practices being discovered on the experiment farms. O'Donnell was appointed to many national positions throughout his agriculture career and served on the Reclamation Commission under Secretary of Interior Franklin K. Lane for President Woodrow Wilson. *Back of photo, from left to right*: W.A. Ryan, comptroller; I.D. O'Donnell, supervisor of irrigation; A.P. Davis, civil engineer; Will R. King, chief counsel. Seated toward front of photo, left to right; F.H. Newell, director; Franklin K. Lane, secretary of interior, (circa 1913). *Courtesy of Sue Delger, great-granddaughter of I.D. O'Donnell.*

5959. Stacking hay, Huntley Experiment Farm. July 1, 1915. D.H.

The new homesteaders were tasked with learning how to turn the virgin soil into land that would produce crops, known as production agriculture. The new farms and commodities were tested by extreme weather conditions and a semiarid climate. Alfalfa was one of the first crops to thrive and be successfully grown for hay and feed. Demonstrations on how to properly stack hay and work with the new equipment being manufactured were done at and by the experiment farms. *Courtesy of the Huntley Project Museum of Irrigated Agriculture Photo Archives.*

Experiment farms on Huntley Project not only demonstrated how to grow successful crops and apply new farm practices but also brought the communities together at yearly events like the much-anticipated annual picnic shown here in 1919. *Courtesy of the Huntley Project Museum of Irrigated Agriculture Photo Archives.*

Growing Montana's Agriculture

The Smith-Lever Act of 1914 created the county-by-county extension services we know today. Agriculture education was now available for the farm and ranch operations and home life. Another focus was the youth. What started early as Boys and Girls Club is now the successful 4-H programming offered nationally and internationally. This charter was issued to the Southview Boosters Club of Vida on November 17, 1931. *Courtesy of the Wanderaas Family Photo Collection.*

O'Donnell participated in the early stages and commended the efforts to get the demonstration farm started. By the spring of 1910, three hundred acres had been set aside for the farm, where today's experiment station stands on Huntley Project. Experiments and demonstrations began immediately, and the smaller farms were able to see a difference and by starting to raise dairy cows and grow alfalfa. Soon other crops were introduced, and they were growing successful sugar beet crops.

In 1914, the Smith-Lever Act formalized extension services, establishing the U.S. Department of Agriculture's partnership with the land grant universities to provide higher education in agriculture. Through the state's land grant college, each county in the state would have an extension agent who helped distribute information from the agriculture college and the experiment station to farmers and their families.[158] The 4-H youth program was developed from the extension program and continues as a strong program for youth today.

A History of Montana Agriculture

The Farmer Rancher

Farms experienced more mechanization during the 1920s and 1930s, and information on new production methods were explored at experiment farms. Advanced technology was on more modern equipment. In time, the farmers were able to learn how to grow other commodities besides wheat and diversify their operations by adding irrigation. When possible, farmers grew alfalfa for hay and planted new crops like barley. These additions allowed many grain growers the ability to convert to combination farms, joining crop farming with raising livestock, recognized now as the "farmer rancher."[159]

Montana ranchers were forced to adapt to more efficient and scientific methods after the Depression. The economy left many facing lower prices

Aerial view of the Glasgow Stockyards in its early years. Livestock auction markets began appearing in the state during the 1930s and 1940s, allowing stockmen better opportunities to be involved in marketing their own livestock. The Glasgow Stockyards opened for business in June 1946. Local businessmen and ranchers contributed to the building of the auction, which ran full services selling hogs, sheep, horses and cattle. *Courtesy of Mark Nielsen.*

and the need to discover new ways to market their products. Ranchers sold their livestock to Midwest feedlots for finishing instead of keeping them on their own pastures. The breakup of the great central beef markets in Chicago created local auction markets. By the 1950s, more than half of the state's cattle were selling through Montana's own auction rings.[160]

12
THE LEGACY CONTINUES

ARTHUR H. LANGMAN AND THE BILLINGS LIVESTOCK COMMISSION

From Grand Island, Nebraska, to Billings, Montana

*A Tribute to Arthur H. Langman,
the Founder of Billings Livestock Commission Company*

The romance of the old west will never die, and much credit must be given the western horse for keeping this romance alive.
W.J. "Bill" Hagen, Editor of Bit and Spur, "News of Horses and Horsemen" weekly column, Western Livestock Reporter, *June 1943*

At one time, the world's horse buyers, traders and sellers came to Grand Island, Nebraska. But on Sunday, December 12, 2004, at 610 East Fourth Street, the end of an era unfolded. The last yard gate closed, the auctioneer's rhythmic cadence no longer resonated through the Bradstreet Livestock Commission sale ring and the last horse sale consignor to buyer transaction was in the books. The descendants of Thomas E. Bradstreet said goodbye to not only a business but also a lifestyle and history that their great-grandfather created in central Nebraska over a century earlier.

The economic impact generated by Grand Island's horse and mule markets on Fourth Street captured the early days of Nebraska's role in the worldwide horse industry. Besides Bradstreet's commission, many others would join in opening horse, mule and livestock markets. In 1924, the Grand Island Horse & Mule Market became known as the world's largest. Inconceivable that the world's largest anything could be completely gone—especially livestock markets that spanned Grand Island's cityscape and crowded Fourth Street south of the Union Pacific Railroad tracks and east of the Chicago, Burlington and Quincy Railroad line near North Plum Street. Today, the only semblance of the era is the iconic Black Stallion horse sculpture that graced the top of the markets and now resides at the nearby Stuhr Museum.

Through the stories, memories and careers of one of the Grand Island's Horse & Mule Market's co-founders, Arthur H. Langman—a bookkeeper turned horse buyer—the legacy of Grand Island's horse market continues today, more than seven hundred miles northwest in Billings, Montana.

The Origin of the Horse and Horse Sale Markets in the West

"Montana Had First Horse" stated a front-page headline of the Tuesday, April 1, 1930 *Billings Gazette* after it was reported that Louis Penwell, a "widely known sheepman" and amateur paleontologist, presented to the Billings Rotarians noon audience.[161] Penwell delivered a paper based on his own understanding of prehistoric Montana and proclaimed that science proved that ancestors of the modern-day horse had their origin in southeastern Montana, a time when Billings and the surrounding territory was thought to be a semitropical jungle.

Prior to its statehood on November 8, 1889, and from 1854 to 1861, Montana was part of the Nebraska Territory. Historians date the first horses in North America to Nebraska about 380 years ago. Brought by Spanish explorers, horses mobilized the Plains Indians, as they used them to hunt, gain economic dominance and prominence, wage war and defend themselves.[162] Although horses were few among early settlers, the revolution in agriculture technology between 1820 and 1870 created a demand for larger, stronger horses to power new farm equipment.[163]

The year after Nebraska achieved statehood in 1867, Union Pacific railroad surveyors platted a town in the Platte River Valley called Grand

Island. That same year, the nation's first veterinary college opened at Cornell University in New York, and farmers became more educated about the care, feeding and breeding of horses.[164] In 1884, the Burlington line opened up from Aurora to Grand Island, Nebraska, and later extended to Billings, Montana.

Before the Homestead Act of 1862, the average American farm was about 100 acres. While oxen and light horses were suitable for tilling the farmland in the East where there was adequate rainfall, a stronger power source was needed to work the virgin soil of the high plains. As farm equipment advanced, so did the demand for draft horses to power them. Chauncey North and William Clarence Robinson acquired a livery stable from the Scudder brothers and opened a sales barn in Cairo, Nebraska, about fifteen miles northwest of Grand Island on main line of B. & M. The North & Robinson Company eventually gained a statewide and national reputation for traveling to France and Belgium to import and breed quality Belgian, Percheron and French draft horses at their 340-acre ranch. By 1900, there were more than twenty-seven thousand purebred Belgians, Clydesdales, Percherons, Shires and Suffolk Punches in the United States. The infusion of these new bloodlines helped increase the average horse size from about 1,200 to 1,500 pounds.[165]

Business and livestock opportunities did not go unnoticed by thirty-seven-year-old Thomas E. Bradstreet. Before moving to Grand Island in 1902 from Sioux City, Iowa, Bradstreet found opportunity in the dairy farming and cattle feeding industries. His first shipment of livestock consisted of eight carloads and later a full train load of twenty-one cars. Stopping in Grand Island for feeding, Bradstreet explored area maps and quickly understood that the city could be one of the principal horse markets of the West. Many, like the North & Robinson Company, had been drawn to the prosperous agricultural confluence of the Wood and Platte Rivers, a relatively new settlement on an area the French traders called La Grande Isle.

In early 1903, with wife Luella and young sons Archie and Deo, Bradstreet moved his family to Hall County, Nebraska, and established his presence in Grand Island. With some assistance of the Union Stock Yards management, Bradstreet's career in the commission horse market business began on June 23, 1903, when he held his first sale at the Union Pacific Stockyards, the site of the Cow Palace and the first livestock auction in Grand Island.

Within six months, Bradstreet sold 4,000 head of horses. Every year, the numbers increased. In 1904, 7,184 head sold, and by 1905, the number of horses sold had doubled, to 8,112. Bradstreet took on a partner, Jess

Clemens, and the newly formed and growing business, Bradstreet & Clemens Co., moved its operation from near the Union Stock Yards to Grand Island's East Fourth Street. Bradstreet began to buy property to make room for more pens. Like the earlier French settlers, Bradstreet recognized the area as the "crosshairs" of western business. The scope of Grand Island's magnetism became undeniable when studying the distance the horses were being shipped from the Grand Island markets. The Burlington Railroad shipped horses in from Oregon, Washington, Idaho, Montana, western South Dakota and Nebraska. In addition, the Union Pacific Railroad system brought horses from Oregon, Nevada, Idaho, Utah, Colorado, Wyoming, Arizona and New Mexico.

As business gained momentum, so did the news coverage of the sales. Editors August F. Buechler and Robert J. Barr of the *History of Hall County Nebraska* wrote of Bradstreet:

> *When he went into a line of business with which he was familiar and for which he was well equipped, since 1903 no resident of Hall County having prospered more substantially in the horse business.... The horse and mule market was the industry that has probably done more than any other one industry ever represented in Grand Island to spread the name of this city over the entire world.*

In 1910, Bradstreet & Clemens Co. built two barns—one to accommodate twenty-five carloads of horses and one to hold fifty carloads with outside pens opposite the barns to carry another fifty carloads.[166] As the business grew, Bradstreet required and acquired the talents of a young bookkeeper, Arthur Henry Langman.

LIFE BEGINS FOR ARTHUR HENRY LANGMAN

Born March 20, 1882, in Grand Island, Arthur was the youngest son and child of Fred (Fritz) and Magreta Langman. He was named after Chester Alan Arthur, an American attorney and politician who served as the twenty-first president of the United States, from 1881 to 1885. Soon after Arthur's first birthday, a "shuddering mishap"[167] on May 25, 1883, claimed the life of his mother. In these days, before waters from the winter melt of the Rocky Mountains were dammed, or bottled up, for irrigation purposes, the spring pilgrimages of the Platte River became a continual source of worry for Nebraska homesteaders and others who had settled near her banks.[168] After most winters of snow, sleet and ice in the far west, the spring thaw would race through the Platte River and slowly subside into smaller outlets.

But then came the spring of 1883. Rains swelled the swift-moving water over the banks and into the Langman family homestead. Fred was away at the family's Prairie Creek farm northwest of Grand Island. Meanwhile, pools of water entered Magreta's kitchen and engulfed the out barns, damaging large quantities of hay and separating mother cows from their calves. Seeing the Langmans' hired man, Hans, across the opposite bank, Magreta quickly gathered her young children and placed them on the highest spot away from the river she could find. Turning her attention back to Hans, she trudged back toward the river and to a small boat and ferry, grabbing the attached rope to maneuver herself to the other side. Then, something broke.[169] The boat bounced and capsized, tossing Magreta into the plunging waters and sweeping her helpless body downstream. It was months later when someone found the skeleton of a human body on a sandbank near Columbus, Nebraska. On the third finger of the hand was a ring bearing the inscription "Fritz to Magreta."[170]

Bringing her remains back to Grand Island, Fred was able to inter her there and begin the reality of caring for his three motherless children. Margaret (Magreta) Catherine Rief Langman was only twenty-eight years old at the time of her death. She left behind her husband, who was born in Parchin, Mecklinberg, Germany, on May 2, 1850, and three children: Frederick (eight), Caroline (four) and Arthur, the youngest at one. Fred came to America in 1866.[171] At sixteen years old, he worked for two years in the coal mines of Scranton, Pennsylvania, with his brother John before striking out on his own and moving to Davenport, Iowa. In 1869, Fred became one of the first homesteaders to Hall County, Nebraska, and started farming south of Grand Island near the Platte River. He married Margaret on April 12, 1874. After her passing, Fred married Ida Krueger in 1885. Langman continued to live and work on his homestead until 1908, when he retired to Grand Island and lived at 407 West Koenig Street until his death on December 27, 1924.

Arthur spent the rest of his childhood working on the eighty-acre Platte River farm with his father and siblings. His brother, Fred C., served as the clerk for the Hall County Court for ten years before moving west to Potter, Nebraska, and opening a garage. Sister Caroline married J.L. Converse, a traveling salesman. The two lived in St. Louis before he died, and she moved to South Pasadena, California. Caroline became a writer and founded the Pasadena Branch of the National League of American Pen-Women, among numerous other clubs. In her 1949 book *I Remember Papa*, she assembled and recounted the early trials and hardships that early Nebraska pioneers, like her father, endured when they came to America and built up the communities in which they lived. She grew up with an unforgettable love for the prairies.[172] She described her younger brother, Arthur, as growing up "to become a good looking chap," who earned money by helping neighbors with chores so he could acquire a horse and buggy.

Arthur H. Langman Becomes Co-Founder of the Grand Island Horse & Mule Market

After leaving the Platte River family farm, Arthur H. Langman delivered groceries for Frank Olson's store. He met folks who were engaged in selling horses and found himself entering the business. The November 9, 1906 *Grand Island Independent* included this announcement: "Arthur Langman will resign his post at Olsen's [*sic*] store and will take a position with Bradstreet & Clemens at their large sales stable now being erected. He will attend to the bookkeeping, etc., of the firm."

Within a month, Langman was making trips to Cairo to buy horses from the North & Robinson sales and bringing his purchases to the Bradstreet & Clemens market in Grand Island.[173] By 1907, North & Robinson had moved its sale barn from Cairo to Grand Island, across from the Bradstreet & Clemens Company barns. With his business partner and bookkeeper in place, Thomas Bradstreet and the firm of Bradstreet & Clemens continued to increase its sale numbers and barns. Bradstreet capitalized on its horse dealing potential by establishing six independent commercial sales operations on East Fourth Street. From 1906 to 1912, they sold 82,236 head of horse and mules, making their average nearly 12,000 head per year. The barns were built between East Fourth Street and the Union Pacific Railroad tracks, with loading chutes accessible for ease of shipments. Some barns could hold twenty-five carloads of horses or more.[174] With the rapid growth, one can only imagine the amount of bookkeeping required of twenty-five-year-old Arthur Langman—and perhaps an opportunity he saw to manage a horse sale barn of his own.

History of Hall County Nebraska authors August F. Buechler and Robert J. Barr wrote that Langman's contribution to his business interest and success could be credited to his native Grand Island for his upbringing, schooling and residence.[175] "The business success that has made his name so widely known, has been secured by persistently following an industrious path in a common sense way, making use of the practical talents that nature bestowed and with good judgment never assuming responsibilities too heavy to carry."[176]

With good horse sale managers, marketers, bookkeepers and rail transportation to and from Grand Island, questions remain: Where did all these horses come from? Who were all the people buying these horses, and where did they come from?

For starters, U.S. Army Cavalry officers were detached periodically to purchase horses for their respective regiments. At the beginning of the twentieth century, the U.S. Army recognized that securing suitable cavalry and artillery mounts proved difficult to find, as the market for riding horses was in a slump and ranchers had stopped breeding their mares.[177] Both the army and the U.S. Department of Agriculture (USDA) feared that the American riding horse was in danger of disappearing. In 1907, Major General James B. Aleshire, the quartermaster general, established the Army Remount Service to create an efficient breeding program and to train personnel in handling the animals to prepare for future mobilization.

Before World War I, the first remount depot station was established at Fort Reno in central Oklahoma. Other stations that followed included Fort Keogh near Miles City, Montana, and Front Royal in Virginia. Fort Keogh was founded in August 1876, following the Battle of Little Bighorn, and named after Captain Myles Keogh, who was killed during the battle on June 25, 1876. Fort Keogh served as a base for patrols to prevent the Cheyenne and Sioux involved in the battle from escaping to Canada. By 1900, the post had become an army remount station, and infantry troops were withdrawn by 1907. During World War I, Fort Keogh served as a quartermaster's depot, with its remount station receiving more horses than any other station and shipping them over the world.[178] By 1919, the U.S. War Department would establish a quartermaster remount depot at Fort Robinson near Crawford, Nebraska. An estimated ten thousand horses passed through Fort Robinson between 1919 and 1931. Among the stallions would be America's first Triple Crown winner, Sir Barton.[179]

While states like Virginia and Kentucky remained known for their Thoroughbred racers, horsemen east of the Mississippi River, like the army's procurement purchasers, looked west for the sturdy, stout and stocky draft horse breeds to bring power to the farm equipment. Buyers and purchasing agents like John "Jack" Washington Torpey traveled to all the major markets in East St. Louis, Illinois; St. Paul, Minnesota; Chicago; Kansas City; Omaha; and Grand Island. From 1902 to 1908, Torpey worked for the Ivins C. Walker's Sales Stables on Valley Forge Road near Norristown, Pennsylvania, searching for well-bred working horses and cattle. Torpey had transitioned into the horse and cattle buying business after a favorable career as a steeplechase jockey.

Torpey was born in Delaware County, Pennsylvania, near Philadelphia, on February 22, 1870, the third of five sons to William and Ellen (Lee) Torpey. His parents were natives of Ireland who were brought to the United States as children.[180] Torpey attended and completed public school in nearby Radnor, Pennsylvania, before becoming an expert horseman and jumper.

Two years after working for the Ivins C. Walker's Sales Stables, Torpey's travels west eventually led him to Columbus, Nebraska, where he met and entered into a partnership with A.C. Scott. The pair opened the Green Front sales and feed yard there in 1910 before relocating sixty miles southeast to Grand Island, where other livery stables, feed stores, horse dealers, blacksmiths and harness shops opened near the Union Pacific Stockyards and along East Third, Fourth and Fifth Streets.

Arthur Langman left his bookkeeping position at the Bradstreet & Clemens Co. by March 18, 1912, when he was in a partnership with the I.C. Gallup Horse and Mule Co.[181] But by November 1912, a newly formed team of W.S. Fletcher of Loup City, Arthur H. Langman (president), John W. Torpey (secretary-treasurer) and A.C. Scott (vice-president) founded and opened the Grand Island Horse & Mule Market at 607 East Fourth Street. William I. Blain served as auctioneer. However, Fletcher only appears on early advertisements in 1913, and a Grand Island Horse & Mule Co. letterhead lists Arthur H. Langman, A.C. Scott and John Torpey as proprietors.

In 1910, Langman married Daisy Heffeifinger, and by February 1914, he seems to have left his position at the Grand Island Horse & Mule Market Co. to operate individually back at Bradstreet & Clemens and to pursue other business opportunities in Grand Island leading up to the start of World War I. Daisy and Arthur Langman had one son, Arthur Jerome "Jerry," who was born August 24, 1918. Jerry was baptized on January 25, 1919, at Grand Island's St. Stephen's Episcopal Church at 422 West Second Street.

By 1919 or the early 1920s, the Arthur H. Langman family had moved farther west toward Denver, Colorado.[182] The Langmans may have lived

Portrait of Arthur H. Langman, co-founder of the Grand Island Horse and Mule Market, Grand Island, Nebraska (1912), and founder of Billings Livestock Commission, Billings, Montana (1934). Circa 1930s. *Courtesy of the Scott Langman (grandson of Arthur H. Langman) family, Billings, Montana.*

in Sidney and Scottsbluff, Nebraska, for a brief period before moving to Colorado. An advertisement for the Colorado Horse & Mule Commission Company in the 1924 *National Wool Grower* lists A.H. Langman as vice-president. "Over one million ($1,000,000) dollars' worth of horses and mules sold in twelve months. Expect to sell more range horses this summer than any commission firm in the world. Outlook is good for very satisfactory prices. Your consignments get protection here."[183] The commission firm was located at the Union Stock Yard in Denver, Colorado.

Nine years later, another advertisement for the Colorado Horse & Mule Commission Company/Denver Union Stock Yards in the *Denver Daily Record Stockman* 1933 Stock Show Edition lists Ray Hayes, C.M. Powell and A.H. Langman as proprietors. "Your business handled promptly and efficiently on a commission basis by a firm that has made Denver the greatest horse market in the United States."[184]

The first two livestock auctions in Colorado were established in 1931.[185] Arthur Langman became president of the Walker-Langman Land & Live

Stock Company of Elbert County, before moving to Billings, Montana, in early 1934 and opening Billings Livestock Commission Company. From its original facilities on First Avenue North near the Northern Pacific rail line, Billings Livestock Commission Company was established as a horse and mule auction. Buyers as far as New York State would come to Billings to purchase their livestock and ship the animals back east on rail cars. It is noted in a Farm Credit Administration Bulletin from 1939 that "the first auction in Montana was started in Billings in 1934." In 1937, only three auctions were active and all operated by one corporation, with much of their livestock coming from distances greater than one hundred miles.[186]

Billings Livestock Commission (BLS)—An Author's Historical Discovery

Nowhere in the United States can they [buyers] *find the numbers or the quality they are able to see each and every month in Billings.*
—*William "Bill" Parker, who managed the Billings Livestock Horse Sales with his wife, Jann, from 1998 until his death in June 2016*

Today, a sign above the Billing Livestock Commission auction block states, "TRUE PRICE DISCOVERY STARTS HERE." One learns quickly at BLS horse and cattle sales that facts trump opinions every sale day. If consignors want to know the true market value of their livestock—not what they think the animal is worth but what buyers attending are willing to pay—then they will soon discover and learn by placing their commodity in the sale ring.

Author Jody Lamp discovered and experienced firsthand how working hard to pursue a dream while maintaining high morals and values can pay off. Before founding and opening her Lamp Public Relations & Marketing business office at Billings Livestock Commission in June 2009, Lamp became acquainted and friends with Jann and Bill Parker, Patrick Goggins and Goggins's family members, who currently own and manage Billings Livestock Commission near Lockwood at 2448 North Frontage Road and the Public Auction Yards (PAYS) at 1802 Minnesota Avenue. From 2007 to 2010, Lamp worked on the auction block for the weekly BLS cattle sales and as the clerk for the monthly BLS horse sales. And she occasionally clerked a bull sale and penned cattle sales at **PAYS**.

Jann and Bill Parker, who were family friends from church, were the BLS horse sale managers when Lamp learned they were looking for help at their monthly horse sales. Lamp agreed to help the Parkers and fill in as the horse sale clerk for a few months until they could find more help. For Lamp, a few months turned into three years. There was something about Billings Livestock and working on the block that resonated within Lamp's spirit and her deeply rooted agriculture upbringing in Nebraska. The place felt like home. The people felt like family. And the perspective from the horse sale clerk's high-top seat gave Lamp a vantage point equal to the auctioneers and pedigree announcers. Every month, with the exception of December, thousands converged to Billings Livestock Commission, where they crowded into the seats surrounding the sales ring to listen to the rhythmic cadence of world champion auctioneers sell their horses or mules, shop for others or do both. With the Parkers at the helm since September 1998, the two-to-three-day horse sales consistently averaged five hundred to one thousand head of grade, registered, loose or catalogued horses and mules every month to continue BLS's reputation as the "Horse Selling Capital of Western America."

Lamp developed a natural and seemingly magnetic curiosity for the history and founding of BLS. In her spare and not-so-extra time, she sought to find out more than what can be seen on the BLS company website's "About Us" section:

> *Back in 1934, the Wolff* [sic] *Brothers left Denver, Colorado and landed in Billings, Mont., striking up a partnership with the late Arthur "Art" Langman. Originally, they created a horse and mule auction and later added cows and bulls. Located on First Avenue North, they leased facilities from the Northern Pacific Railroad and set up shop.*
>
> *Billings LiveStock Commission was the hub market as the Wolff-Langman Partnership developed markets also in Great Falls and Miles City, MT. Along with auctioneer, Norman G. Warsinske and cattle buyer, Lyle Devine, they created a livestock merchandising endeavor that has never been equaled.*
>
> *In later years Art's son, the late A.J. "Jerry" Langman and Ralph Cunningham, along with a fieldman by the name of Conrad Burns, now Montana's US Senator, continued the BLS tradition. During the late 1970's Scott Langman, Jerry's son, became the third generation operator and moved the Billings Live Stock Commission to its present location on the North Frontage Road east of Billings.*

> *In 1984 Scott sold the business to Patrick K. Goggins, who operated it for some years. Pat sold it to Jack McGuinness, who operated it for sixteen years. BLS was purchased in 2003 by Goggins, who totally rebuilt the stockyards, revamping it from head to tail.*
>
> *BLS sells cattle every week on Thursdays for all classes. The 4th weekend of each month, BLS becomes the "Horse Selling Capital of Western America." 500 to 1100 head sell on any given weekend. The Northern Livestock Video Auction base operation is also located at BLS. Several video sales a year are staged to an international market.*
>
> *One of the oldest, continuous livestock auctions in America, BLS is a great tribute to its pioneer founders.*[187]

In a quest for a timeline of when and how Art Langman came to Billings and opened Billings Livestock Commission, Lamp began searching for the earliest news article or reference she could find.

> *August 3, 1935*
>
> *"35 Carloads of Horses Sold At Billings Yards," stated the headline in the* Billings Gazette.[188] *Reportedly, 212 head of horses were sold that Friday, August 2, beginning at 8:00 a.m. and ending at 10:00 p.m. It was the largest number ever sold at the new stockyards at First Avenue North, which could now accommodate 3,000 head of stock, and another sale was already planned for Friday, August 16, 1935.*

By Friday, December 13, 1935, the *Billings Gazette* carried a headline and year-end recap story donning its front page: "Cattle, Horse Sales at Yards Here Are Heavy. More Than 3,000 Carloads Shipped During Year, About 100 Trains of Livestock."[189]

> *In the announcement that more than 3,000 carloads of cattle were shipped from Billings to points south and east over the Burlington and Northern Pacific railroads, Langman pointed out that before the commission company began doing business here the average annual shipment of stock from this point was 69 cars. The company has been doing business here not quite two years. Three thousand carloads from, roughly, 100 trains.*
>
> *An important horse sale to be held Monday, December 30, was announced Thursday night by Mr. Langman. Numerous telegraphic*

inquiries relative to this sale have already been received from buyers at distant points, including Maryland and Tennessee....

While cattle and horse sales have usually, in the past, alternated every other week, the commission company has made plans to hold both cattle and horse sales each week during the first three months of 1936....

Mr. Langman left Thursday night for Kansas City, Mo., on company business.[190]

On December 31, 1935, Langman's Billings Livestock Commission company closed a year "which marked an important milestone in the city's livestock history."[191] By year end, the books showed 71,132 head of cattle and 16,847 head of horses, with gross business receipts "in the neighborhood of $4,000,000,"[192] a number equivalent to more than $74 million in purchasing power eighty-five years later, according to the 2020 Bureau of Labor Statistics consumer price index inflation calculator.[193] Less than two years in business, Arthur H. Langman's Billings Livestock Commission was recognized as one of the largest enterprises of its type in the West.

A March 22, 1936 article in the *Billings Gazette* stated the following:

One of Eastern Montana's biggest industries, the Billings Livestock Commission company, will celebrate its second anniversary this week with a series of livestock sales that are expected to draw to Billings hundreds of Montana and out-of-state buyers....

The horse sale will be broadcast over the radio from 9 to 10 a.m. on Monday....

That the livestock commission firm has become an integral part of the community's life is shown by its telegraph and telephone bill of $5,380. Labor and salaries during the period total $53,000. Advertising has amounted to $19,784.

The average atendance [sic] *at the sales is estimated at 600 persons per week.*[194]

In an effort to pinpoint the company's inaugural sale, Cindy Newman suggested, "Why don't you just go ask Scott Langman? He was Art Langman's grandson and still lives in the area." Newman, an assistant supervisor at the BLS brand inspector's office, explained that she had worked at the current BLS building location on the North Frontage road since 1976, the year when its president, Scott Langman, moved the markets from its original location, 1202 First Avenue North near downtown Billings.

Langman purchased 34 acres of land near Lockwood from Billings businessman and real estate developer Richard Popelka. Moving the stockyards and building a new, modern sales barn facility worth $750,000 made BLS the largest in Montana.[195] The Lockwood site featured new lagoons and its own Environmental Impact Statement to meet all Environmental Protection Agency (EPA) standards; 10 acres of parking, dwarfing the downtown location; and modular checkerboard-pattern pens to hold 7,000 head of cattle or over 10,000 head of sheep all designed to ease cleanup. Inside, the new sales ring would seat 450 buyers and feature a $35,000 ring scale that would record average and gross weights, head count and average price per head. Also featured inside are several offices and a restaurant with 1,500 square feet of seating space.[196]

A *Billings Gazette* article, dated February 11, 1934, stated that Arthur H. Langman, a prominent Denver livestock commission broker, would be

A typical scene of an Arthur H. Langman Billings Livestock Commission horse sale on First Avenue North, circa 1945. This bird's-eye view reveals the outdoor sales area pointing west toward downtown Billings. Spectators at least a dozen rows deep fill the seats, surrounding the consigner, whose path of impression narrowed to that of an alleyway. Young lads sit perched on rafters just to elevate their height over the buyers standing below. And most notably, a glimpse into the Magic City's bygone years when a distant Northern Hotel towered over any other structure in downtown Billings. *Courtesy of the Scott Langman (grandson of Arthur H. Langman) family, Billings, Montana.*

A History of Montana Agriculture

A postcard inscription describes "23 Cars of Horses on special Train" leaving the BLS on March 2, 1935. It was recorded that before Arthur H. Langman commenced operating Billings Livestock Commission, shipments of about 400 cars of livestock originated annually in Billings. During 1935, train carload shipments in and out of Billings were slightly more than 3,500 carloads. *Courtesy of the Scott Langman (grandson of Arthur H. Langman) family, Billings, Montana.*

opening a livestock auction market on First Avenue North in Billings on or about March 1, 1934. Langman made the announcement via telegram to the city's newspaper the day before with the full disclosure that he would lease about eight acres of pens and yard from Fred C. Pierce, owner of Pierce Packing, for his venture.

> *Mr. Langman stated his concern is to be known as the Billings Livestock Commission Company and will be capitalized at $20,000. The corporation is to be controlled by the commission man and a group of associates it was said here.*
>
> *The company, it was stated, plans to buy and sell horses, mules, cattle, hogs and sheep both on a commission and auction basis, providing a stable livestock market for this community.*[197]

In a *Billings Gazette* article, "Old Dobbin Worth More Than in '33, Auction Man Avers," dated Thursday, March 22, 1934, the writer states:

Horses today are bringing 50 percent more than they did on the market a year ago. Authority for this statement is A.H. Langman, president of the recently opened Billings Livestock Commission company, who Wednesday based his opinion on prices posted at the concern's inaugural auction, Monday.

And now, within these pages, told for the first time in one complete manuscript, the legacy of Arthur H. Langman, who moved to Montana with his wife, Daisy, and son, Jerry, in early 1934; founded the first livestock market in the state at 1202 First Avenue North in Billings; held the inaugural sale on March 19, 1934, the day before his fifty-second birthday; and served as its president until his death on September 9, 1952.

From his birthplace of Grand Island, Nebraska, where he co-founded the Grand Island Horse & Mule Market, to the city and state of his final resting place—Billings, Montana—the pioneering spirit and legacy of America's horse and mule markets continues to thrive.

Langman's Sales, Tales and Influence Through the Years

In its infancy, the Billings Livestock Commission Company was primarily a horse and mule market, selling 1,200 horses weekly. But after a few short months, Langman realized the need for a full-scale livestock auction market in Montana and believed the area ranchers and farmers were entitled to an active cattle market in their own community. Langman launched an extensive advertising campaign and direct-marketing program to attract buyers to Billings. It was reported on the front page in the June 1, 1936 *Billings Gazette* that "the commission company contacted in the last 10 days through telephone, telegraph and mail some 12,000 prospective buyers."[198] The investment proved successful, as buyers realized the vast savings on freight and shrink by consigning with Langman's local market. And certainly, the spirit of competitive bidding on each and every head of livestock ensured top prices.

Less than two years after the grand opening of Billings Livestock Commission Company, Langman was set to open his second livestock market in Great Falls, Montana. Plans for the construction of a sales pavilion and livestock yards were drafted and presented by Langman of Billings and Julius Wolf of Denver reported the *Great Falls Tribune* on

Saturday, February 8, 1936. Local businessmen noted that as soon as weather permitted, ground would break on the ten-acre site north of the North Montana fairgrounds with an estimated cost of $20,000. And Langman cited the success of Billings Livestock Commission Company as an example of how the stockmen throughout Montana's northern ranges would benefit from a stockyard in their area and how it would stimulate their economy.[199]

> "We took over the Billings yards two years ago, at which time it was inactive. At our first sale, we had only 280 head of cattle to offer. Business has increased steadily, and at one sale a few months ago, we disposed of 4,700 head, our largest day to date. In the last nine months, the yard sold 70,200 head of cattle. Last we sold a little more than 25,000 horses.
>
> Before we commenced operating, shipments of about 400 cars of livestock originated annually in Billings. Carload shipments in an out of Billings during 1935 were slightly more than 3,500 carloads....
>
> At sales such as are conducted at the Billings yard, and such as will be conducted here, all buying is done by men who know their business. They know quality when they see it, and will pay for it. As soon as it becomes generally realized that the man who goes home with the fattest checks from these sales is the man who has offered a lot of quality stuff, rather than the man whose offerings have been mediocre, I predict that you will see a marked increase in the quality of stock on northern Montana ranges," he [Art Langman] observed.[200]

As drought conditions persisted through the mid-1930s, western ranchers sought auctions to bring their cattle off the range. By 1936, Langman's Billings Livestock Commission Company and partner William "Billy" Wolf* responded by establishing livestock auction markets in Great Falls and Miles City, Montana. Langman now owned the only three markets in Montana. Another concerted move to ensure the cattle and hog auction business was Langman and Company efforts to attract national meatpacking companies and buyers to the area. Armour and Co., Swift and Co. and Wilson and Co. all stationed permanent buyers at the auction; in addition, there were buyers from two local packers, Midland Empire Packing Co. and Pierce Packing Co.

---

*The authors acknowledge that William "Billy" Wolf and the Wolff brothers are referring to the same person/family. However, the last name is spelled differently in multiple references.

Back in Grand Island, Nebraska, at the Grand Island Commission Company, Langman's friend and former business partner John "Jack" Torpey Sr. had purchased Grand Island's Blain Livestock Commission Company in 1931 and organized the Grand Island Livestock Commission Company, consolidating the horse and mule, cattle and hog business under one institution. Cattle, hogs and sheep sales were held every Monday and horse and mule sales held every Wednesday. Torpey's leadership in the horse and livestock industries made a national and worldwide impact; certainly, his connection with Langman influenced the early days of Billings Livestock Commission Company.

John W. "Jack" Torpey Jr. was born on June 26, 1919, in Grand Island, Nebraska, to Torpey Sr. and his wife, Lillian Hinz Torpey. Seven years before Jack's birth, his father founded and organized the Grand Island Horse and Mule Company along with business partners Arthur H. Langman and A.C. Scott. By November 1924, when Jack Jr. was five years old, the Grand Island Horse & Mule Company surpassed the largest market in East St. Louis, Illinois, making Grand Island, Nebraska, home of the world's largest horse and mule market.

Jack Jr. was raised at the livestock markets on Fourth Street in Grand Island alongside his father when horses and mules were still part of everyday life. Many believed with the advent of the "automobile, bus, fast train and airplane" the "old dobbin" would be headed to its last roundup.[201] But horses from "slow-moving drafts to the well-groomed riding mounts" continued to bring top dollar in Grand Island and later in Billings, Montana, too. On multiple research trips to Grand Island, the authors' heard a story that when Jack Torpey Jr. was fifteen years old, he and his father drove from Grand Island, Nebraska, to Billings, Montana (depending on the route, a seven-hundred-plus-mile trip) to attend the Grand Opening of Billings Livestock Commission Company on First Avenue North to support their good friend—Arthur H. Langman.

Lamp and Dobson also heard in 2015 that Jack Jr., at ninety-six years old, was still alive and buying cattle. They asked his nephew in Grand Island if he could help put them in touch with his uncle, who resided in Long Beach, California, but was on a cattle-buying trip to Great Falls, Montana. It was about a week before Thanksgiving when the authors called and introduced themselves to Torpey Jr. and asked if they could interview him about his memories of the Grand Island Horse and Mule Market and Billings Livestock Commission. Jack agreed but said it would be better for him to visit after the first of the year when he wouldn't be as busy buying cattle. "Be as busy" at age ninety-six? Wow! Jack explained that he was buying twenty-five thousand head of cattle to winter in California and then would transfer the herd to Colorado in the summer.

The authors agreed to call Torpey back after the holidays, but during their brief conversation, he confirmed the story Lamp and Dobson heard. Yes, it was true that he and his father attended the grand opening of Billings Livestock Commission. He also confirmed what the authors suspected—John Torpey Sr. continued to keep in touch with his former business partner, Arthur H. Langman, even after the Langman family moved away from Grand Island. The authors' last words to Jack Torpey Jr. in 2015 were "Happy Thanksgiving….Enjoy your time in Great Falls….Take care and see you soon…after the holidays."

Unfortunately, on December 2, 2015, Jack Torpey Jr. passed away in Santa Cruz, California. His obituary appeared in the *Grand Island Independent* on December 9: "He was a superb horseman, a great judge of cattle and an accomplished fly fisherman. Jack was a wonderful businessman. He was always an optimist, through his many ups and downs, and continued to travel to Nevada, Utah and Montana to the end. He never gave up and he never quit."[202]

After Jack Torpey Jr.'s death, the authors discovered in subsequent research an advertisement for "Two Big Horse Sales" in a June 1934 issue of the *Billings Gazette*. The advertisement provided proof that John "Jack" Torpey Sr. and Arthur H. Langman stayed connected through the horse sales markets. Torpey conducted business in both Nebraska and Montana. As stated in the ad, two big horse sales with 1,000 to 1,500 horses at each sale would be held in in the railroad stockyards in Harlem, Montana, June 28 and 29, and in Miles City, Montana, July 5 and 6. Horses were sold at auction in carload lots, single and pairs and consigned by the Chappel brothers, owners of the famous CBC brand and ranch near Landusky, Montana, with John "Jack" Torpey, sales manager and president of the Grand Island Livestock Commission company of Grand Island, Nebraska, conducting the sale.[203] Torpey acted as the sales manager for the Chappel brothers with another sale advertised for them in Laramie, Wyoming, on August 3–4, 1934.[204]

Back in Grand Island, Torpey Sr. held seventeen horse and mule sales from May 18 through December 28, 1938. Less than two years later, Torpey Sr. died on Monday, April 22, 1940, at age seventy from coronary thrombosis, a heart affliction, at the St. Francis Hospital in Grand Island, Nebraska. Relatives, friends and mourners from about forty states attended Torpey's funeral services at St. Mary's Cathedral in Grand Island.[205] One of Nebraska's best-known men in the horse and cattle business, Torpey was active in the chamber of commerce and served various state livestock and marketing organizations as president. In his later years, he was active in legislation and regulation affecting truck transportation and had a financial interest in the operation and management of a market near Janesville, Wisconsin.[206] He also was one of the principals in a group that was planning the establishment of a packing plant at the old Grand Island Canning Factory.[207]

Langman was fifty-eight years old, and Jack Torpey Jr. was only twenty years old when his father died. World War II was raging in Europe. Germany had invaded Denmark and Norway, and Winston Churchill was about to be appointed prime minister of the United Kingdom. Young Jack attended Creighton University in Omaha before joining the U.S. Army in 1942. The authors speculate that Arthur Langman remained close to Torpey Jr. after his father's passing. Torpey Jr. moved to California from Grand Island, Nebraska, in 1947 after being released from active duty in 1946 with a rank of captain in the Medical Administrative Corps. He became a cattle buyer for Alkali Cattle Co. at the Los Angeles Stockyards, which rivaled Chicago and Omaha as the nation's livestock centers. Later, he went out on his own to form Torpey Cattle Corporation. Over the years, he bought and shipped hundreds of thousands of feeder cattle from all over the West to ranchers and feedlot operators. And for more than sixty-five years, he traveled to Great Falls, Montana, living there through the months of October to January, buying and shipping Montana cattle to various points. Jack Torpey Jr. was known in the Great Falls area as one the last of the "old-timer" cattle buyers.

By September 23, 1940, Langman no longer owned the only livestock markets in Montana. A new stockyard constructed by the Saint Paul Union Stockyards Company of St. Paul, Minnesota, opened near Billings Livestock Commission Co. on Minnesota Avenue called the Union Stockyards Company. The name later changed to Billings Public Stockyards Co. and then to its current name—Public Auction Yards (or PAYS), when Pat Goggins took ownership in 1968. The room for more livestock business proved evident, as horse sale prices and numbers held steady at Billings Livestock Commission Co. through 1942. Buyers from Colorado, Nebraska, Tennessee, Arkansas, North Carolina, Pennsylvania and California continued to frequent the markets, and top mares sold for $110 each.

However, Langman cautioned consignors that "inferior kinds and ragged colts are not sought for and prices are low."[208] Registered stallions continued to be in high demand in the western horse sale markets as U.S. Army Remount Service Saddle stock. By June 2, 1942, Billings Livestock

Commission Co. had sold 3,654 horses,[209] with a year-ending total of 6,591 horses and a continued reputation as the leading horse market in the West.[210] On Tuesday, October 27, 1942, the *Western Livestock Reporter* featured a front-page article with photos of the Billings Livestock Commission Company cattle sale results. It was reported that "$14.95 pen of calves consigned to sale by Rigler Bros. (of Corwin Springs, Montana) and purchased by Wm. Walker of Norristown, Pennsylvania."[211]

> *As mentioned previously, John Washington Torpey Sr. was born in Delaware County, Pennsylvania, near Philadelphia on February 22, 1870. From 1902 to 1908, Torpey worked for the Ivins C. Walker's Sales Stables on Valley Forge Road near Norristown, Pennsylvania, searching for well-bred working horses and cattle.*[212] *Ivins C. Walker died on Thursday, July 27, 1939. William W. Walker, a presumed relative, served as pallbearer.*[213]

As a salutation to end 1942, Langman's business occupied nearly the entire third page of the December 22, 1942 edition of the *Western Livestock Reporter*. A "1943 Season's Greetings" three-quarter page advertisement from the Billings Livestock Commission Company and the Wolf-Langman Livestock Commission Co. of Great Falls, Montana, was signed by "Art Langman" and delivered the following message:

> *It is with pride we feel that you have been pleased by the sales of your livestock through our auction markets, which was evidenced by your loyal patronage....Be assured that all our interests and efforts are for the best prices and to always conduct a market that you too can refer to with the same spirit and good will toward us, as we have toward you.*
>
> *Don't think that we do not realize the fact that your co-operation has helped build this the nation's finest market system which is the envy of all competition far and near....For this we humbly thank you.*
>
> *But...in all these troublesome times to you as well as to ourselves who have dear ones in the service, let us all take time out to "SALUTE" these noble boys and say a little prayer for their safety and speedy return.*
>
> *And remember...the few sacrifices we make may mean a life to them and we should bow our heads in shame if we complain. Let us buy bonds. Let us ration any and everything that will help. We are the people behind the man behind the gun. Let us all help.*
>
> *Art Langman*

As the United States and its Allies gained momentum against Japanese and German forces from 1943 to 1944, buyers from fifteen different states purchased cattle, hogs, sheep and horses at the BLS, during the first week of February 1943. Horses moved out to buyers from North Carolina, South Carolina, Virginia, Tennessee, Missouri, Nebraska, Minnesota, Arkansas and Georgia.

"The horse market has been the best in the past three to four years," said Mr. Langman. "The demand is for the good broke, work and saddle horses."[214]

In the April 27, 1943 issue of *Western Livestock Reporter*, a new column by W.J. "Bill" Hagen, editor of *Bit and Spur*, called "News of Horses and Horsemen" shared the following:

> *It is hoped that this Horse Feature will find a ready response among you readers who are interested in horses.…Horse interest everywhere has increased, as more people have found them to be the perfect answer to the inborn spirit of all Americans' natural love for animals.…At a recent saddle horse sale held in Billings, bidding was brisk on every good horse that was shown in the ring. Buyers attended from ten states. They were not slow in recognizing a good horse, competition was keen and the results of this sale proves* [sic] *that it pays to raise quality horses, the kind that will find a ready buyer at a good price.*[215]

Langman's sales success and increased livestock numbers required improvements to the stockyards by May 1943. New covered sheds with concrete floors were added, and alleyways resurfaced. In 1942 alone, a total of 1,485 carloads of all classes of livestock were shipped out by rail to thirty-three states from the Billings Livestock Commission Company from 7,684 individual farmer and rancher consigners. It was reported that from 1937 to 1942, a five-year total of $19,172,639 worth of livestock had been sold.[216]

By October 1943, Lyle Devine, formerly of Billings and the Billings Livestock Commission Co. and Wolf-Langman Co. of Great Falls, had acquired the Miles City Commission firm.[217] Devine's wide acquaintance from both the Corn Belt and West Coast was noted as being a "big factor in making the Miles City market one of the foremost in the west."[218]

Three months after World War II ended, A.J. "Jerry" Langman, Daisy and Art's only son, was officially released from active duty on December 10, 1945. After four years serving with the U.S. Air Force, Jerry, twenty-seven, returned to Billings and assumed many of the duties connected with operating

A History of Montana Agriculture

A sign from "Billings Live Stock Commission" that still existed inside the sale ring area inside the former Metro Park property at 1140 First Avenue when coauthor Jody L. Lamp was conducting her history of Billings Livestock Commission research in 2009. *Photo by Jody L. Lamp.*

the Billings Livestock Commission Company. Two years earlier, Verda Lucille "Pat" Sanback, a Broadview native, had taken a position at the Billings Livestock Commission Company as Art Langman's administrative assistant. Pat was twenty-three years old, had already attended Billings Business College and worked as a secretary at Lockheed Aircraft Company in Los Angeles. When she returned to Billings in 1943, she worked at Billings Livestock for the next five years, providing secretarial services and clerking the cattle and horse sales. Pat was "very fond of Art Langman and in later years related her remembrance of him and his kindness to her with great affection."[219] Pat and Jerry met at Billings Livestock and were married in Billings on November 24, 1945, two days after Thanksgiving. To celebrate America's first Thanksgiving since the victory in World War II, the president issued a proclamation:

Now, Therefore, I, Harry S. Truman, President of the United States of America, in consonance with the joint resolution of Congress approved December 26, 1941, do hereby proclaim Thursday November 22, 1945, as a day of national thanksgiving. May we on that day, in our homes and

Growing Montana's Agriculture

The Billings Livestock Commission Company horse sale catalogue from May 1946. *Courtesy of the Scott Langman (grandson of Arthur H. Langman) family, Billings, Montana.*

> *in our places of worship, individually and as groups, express our humble thanks to Almighty God for the abundance of our blessings and may we on that occasion rededicate ourselves to those high principles of citizenship for which so many splendid Americans have recently given all.*[220]

The year 1945 was ending on a great note. With World War II officially ended, Art Langman was named one of the original directors of the Billings State Bank. Jerry returned home, where his parents resided at 1230 North Thirtieth Street, and started working with his father at Billings Livestock Commission Company and soon would marry Pat, one of Langman's beloved employees. With Jerry's assistance, Arthur Langman would acquire the Wolf brothers' interest in Billings Livestock Commission Co. in 1946.

On Friday, April 2, 1948, the third and last Langman to lead Billings Livestock Commission Co. as its president arrived when Jerry and Pat welcomed their only child, Arthur "Scott" Langman. The year ended with record-breaking numbers in both cattle sold and dollar volume for Langman's firm. What started as a weekly horse sale market in 1934, Billings Livestock Commission Company expanded into regular weekly cattle sales. On December 3, 1948, Arthur Langman reported that all previously yearly records had been shattered with ninety-two thousand cattle sold, valued at $14,500,000.[221]

The Langman Legacy Era

Arthur H. Langman, the founder and patriarch of Billings Livestock Commission Co., died on Tuesday, September 9, 1952, at seventy years old at St. Vincent's Hospital in Billings. That same day, the board of directors of the Billings State Bank met at their regular meeting. A resolution was passed to express their loss in the passing of Langman, who had served as a director of the bank since it was organized. The resolution was signed by O.B. Silvey, president, and a copy was mailed to Daisy Langman, along with a letter dated October 15, 1952. Within the resolution, Silvey stated:

> *Arthur H. Langman was one of the original Directors of Billings State Bank and served in that capacity from the time of the organization and opening of the bank until his death only the day before its seventh anniversary. "Art" was one of the moving spirits in presenting the application for charter and following it thru the many conferences prior to its issuance.*
>
> *During the seven years of operation and growth the bank owed much to his aggressive and progressive efforts in its behalf. Even during his illness and declining health "Art" attended almost every meeting of the Board and added his wisdom, seasoned with his natural humor, to the deliberations.*
>
> *The community will feel the loss of a pioneering, colorful citizen in "Arts" passing, but the memory of his leadership in all phases of livestock producing and marketing will be landmark in the area for some time to come.*
>
> *In recognition of the valuable service Arthur H. Langman has performed as a good citizen of our community and the duties discharged so faithfully and well as a founder and Director of Billings State Bank, an in expression of deepest sympathy to the members of his family, the Board of Directors of Billings State Bank adopts the following resolution:*
>
> Be It Resolved that the Board of Directors of Billings State Bank herby record their deep appreciation of Arthur H. Langman's years of service in helping to establish Billings State Bank as a dependable business institution of the community and their heartfelt sympathy to the members of his family in their irreplaceable loss.
>
> Be It Further Resolved that a copy of this Resolution be entered in the Minutes of the Directors Meeting, and a copy be sent to the family of Arthur H. Langman.
>
> *O.B. Silvey*
> *President*

```
O. B. SILVEY, PRESIDENT                    J. J. MILLS                       J. E. VOGEL, VICE PRESIDENT & CASHIER
F. C. PIERCE, VICE PRESIDENT         CHAIRMAN OF THE BOARD                   A. J. KRINGS, ASSISTANT CASHIER
```

BILLINGS STATE BANK

CAPITAL AND SURPLUS $300,000.00

BILLINGS, MONTANA

RESOLUTION

Board of Directors
Billings State Bank
Billings, Montana
September 9, 1952

In recognition of the passing of Arthur H. Langman this afternoon, the following Resolution was adopted on motion made by Mr. Mills, seconded by Mr. Bateman and unanimously carried:

"Arthur H. Langman was one of the original Directors of Billings State Bank and served in that capacity from the time of the organization and opening of the bank until his death only the day before its seventh anniversary. "Art" was one of the moving spirits in presenting the application for charter and following it thru the many conferences prior to its issuance.

During the seven years of operation and growth the bank owed much to his aggressive and progressive efforts in its behalf. Even during his illness and declining health "Art" attended almost every meeting of the Board and added his wisdom, seasoned with his natural humor, to the deliberations.

The community will feel the loss of a pioneering, colorful citizen in "Arts" passing, but the memory of his leadership in all phases of livestock producing and marketing will be a landmark in this area for some time to come.

In recognition of the valuable service Arthur H. Langman has performed as a good citizen of our community and the duties discharged so faithfully and well as a founder and Director of Billings State Bank, and in expression of deepest sympathy to the members of his family, the Board of Directors of Billings State Bank adopts the following resolution:

On Tuesday, September 9, 1952, the same day that Arthur H. Langman, the founder and patriarch of Billings Livestock Commission Co., died, the board of directors of the Billings State Bank met at their regular meeting and drafted and passed a resolution to express their loss in the passing of Langman, who had served as a director of the bank since it was organized. The resolution was signed by O.B. Silvey, president, and a copy was mailed to Daisy Langman, along with a letter dated October 15, 1952. *Courtesy of the Scott Langman (grandson of Arthur H. Langman) family, Billings, Montana.*

Upon Art's death, Jerry was named president and manager of Billings Livestock Commission Co. In addition, Jerry Langman was elected a member of the board of directors of the Billings State Bank and succeeded his father on the bank's board.[222] Along with Daisy and Pat, the Langman family continued to maintain the livestock yards and manage the family's twenty-two-thousand-acre cattle ranch thirty miles north of Billings.

Other cattlemen, producers and auctioneers to join and partner with Jerry included Ralph Cunningham, R.J. "Bob" Thomas, Tige Thomas and Frank Price. Langman was a member of the Montana Stockgrowers Association and Montana Auction Market Association. He served as president of Billings Livestock Commission until his passing on Saturday, November 8, 1969, at the age of fifty-one. Survivors included his widow, Pat; mother, Daisy; and son, Scott Langman. Only twenty-one years old, the youngest Langman was attending the University of Wyoming and immediately assumed a new role, president of Billings Livestock Commission Co., just as his father had done seventeen years earlier.

The 1970s ushered in years of growth, change and more of the same for Billings Livestock Commission. All classes of livestock sales continued to sell, led by the talent of a farm kid from Missouri turned auctioneer. Conrad Burns, who later became Montana's longest-serving Republican senator in the state's history (1989–2007), traveled Montana as a field representative for *Polled Hereford World* magazine.[223] He turned down a transfer to Iowa to become a cattle auctioneer for the Billings Livestock Commission. Burns became the first manager of the Northern International Livestock Exposition. He started reporting agricultural market news, launched a radio show and later worked as a farm reporter for KULR-TV.

By the end of its fortieth year in 1974, Billings Livestock Commission Co. issued a "Happy Holidays" greeting advertisement to express thanks for its patrons' loyalty and to announce that its president, Scott Langman, would be "taking a leave of absence to complete his education, but feels he is leaving the operation of the business in good hands."

BLS had become known as the "North Yards," as it was north of the railroad tracks. Soon after Pat Goggins purchased Billings Public Stockyards in 1963, now known as PAYS, Billings became the second-largest livestock market in the United States, second in volume only to Amarillo, Texas.[224] Although total gross receipts from Billings Livestock Commission Co. from 1973 to 1974 equaled $52 million, rumors abounded about the future of the company. As Scott Langman searched for available ground to build "the most complete livestock yards and auction facilities between Chicago and the West Coast," he assured the public that the Pierce Corporation had extended the lease on the current yards on First Avenue North. Like his grandfather had done in 1934, Langman turned to the local newspaper to deliver his message and intentions. The advertisement appearing in the *Billings Gazette* on May 24, 1975, stated:

> *Dear Friends of "Billings Live":*
>
> *…As president, I hope to carry on as efficiently as possible what my Grandfather, Art Langman, established 41 years ago, and what my Father, Jerry Langman, so ably helped perpetuate.*
>
> *All of us at "Billings Live" look forward to serving you in the same dependable, reliable manner we have in past.*
>
> *Sincerely,*
> *Scott Langman*

The move to the new Lockwood headquarters was approved in April 1975 by the Montana Board of Livestock, after a public hearing.[225] Although Daisy Langman passed away in June 1973 before seeing her grandson's achievement of continuing the Billings Livestock Commission Co. operations at its new and current facility, Scott Langman's mother, Pat, remarried, and her husband, Horace Wyman, would serve as BLS secretary-treasurer. The new facility would still feature much of the same crew and the original Billings Livestock Commission Co. sign from the downtown location. Touted as the most modern, functional, practical and comfortable auction market in America, BLS management held it first big cattle at 2443 North Frontage Road on January 12, 1976. The first horse sale was held on January 24, and five more cattle sales were scheduled that month for January 15, 19, 22, 26 and 29.

Scott Langman led Billings Livestock Commission into the 1980s. Two months shy of his thirty-fourth birthday and one month before the company's fiftieth anniversary, rumors once again drifted through stockyards—this time of a potential sale of BLS to PAYS owner Pat Goggins. When called and asked by the local farm and ranch editor if the rumors were true, it was reported that Langman's only comment was that before a transaction like that could happen, a public hearing with the state's board of livestock would need to take place. He confirmed that a

Taylor Brown, known as the "Voice of Montana Agriculture," welcomes and speaks to guests attending the 50th Anniversary of Billings Livestock Commission in 1984. Scott Langman, grandson of BLS founder Arthur H. Langman, moved the livestock market from its First Avenue North location to its current site at 2443 North Frontage Road near Lockwood. Touted as the most modern, functional, practical and comfortable auction market in America at the time, BLS management immediately resumed sales after the move and held its first big cattle sale on January 12, 1976. *Photo found by Jody L. Lamp in the archives at Public Auction Yards; courtesy of Bob Cook.*

hearing was scheduled for February 24, 1983. The transaction and transfer of ownership took place.

Twenty-seven years later, Langman expressed his desire to pursue other opportunities at the time as the reason behind selling the BLS to Goggins. From age twenty-one to thirty-three, Langman served as Billings Livestock Commission Company's president, alongside family and close friends who had known and worked with his grandfather and father. By the end December 1983, Langman had joined the investment firm of Piper, Jaffray and Hopwood Inc. Although the Langman era of ownership at BLS officially ended after the third generation, the company remains a testament of success today for the mission, vision and passion of its founding and subsequent leadership in Arthur H. Langman, A.J. "Jerry" Langman and Arthur "Scott" Langman.

In a letter dated July 1, 2015, Scott Langman wrote to the authors:

> *At my current station in life, I have come to realize how much I owe to my father and grandfather. I think of my grandfather Art as a special man. He was mentally a tough man, who lived through some tough times, took big risks, worked hard, was honest, well thought of and successful. He was one of the real pioneers in the Montana livestock industry and was well thought of by many of the livestock producers in our area. I can't begin to tell you the number of old timers who I met while managing the Billings Livestock who told me they "knew my grandfather and how much they thought of him." What better testimonial is there than that?*

Montana's Pioneer Marketplace Magic Lives On

After the sale of BLS to Pat Goggins of PAYS, Bud Knight was named manager. The Custer, South Dakota native earned the Livestock Marketing Association's World Champion Auctioneer title later that year. In 1987, Goggins leased and then sold BLS to Jack McGuinness. Sixteen years later, in 2003, Goggins bought BLS back from McGuinness, and at this time, the yards were completely rebuilt. Goggins retained the routine sales that area livestock producers had grown to know. There was, however, a consolidation of the horse sales, which had been held once a month at both BLS and the PAYS. Since the BLS horse sales had been run successfully by

Bill and Jann Parker with McGuiness since 1998, Goggins made the sound decision to combine the two sales at BLS and keep them under the Parkers' competent supervision.[226]

From September 1998 until the time of Bill's passing on June 23, 2016, the Parkers as a couple and as horse sale managers brought their own distinct business savvy, style and mannerism to the arena, sales barn and auction block. Not one person in North America who has ever attended a Billings Livestock Horse Sale under the Parkers' management couldn't recite their motto: "At BLS—We love horses!" Each horse sale was dedicated to a featured discipline that buyers and consignors alike came to know and depend on when planning their monthly trips to Billings. For example: January, Winter Special/Tack Sale; February, "Sons and Daughters" of top money-earning horses from National Finals Rodeo qualifiers; March, Foundation-bred, Outfitters; April, Spring Catalog and Rope Horse Special; and May, the BLS Cow Country Classic.

"This sale has been a feature sale on the calendar since 1934, I can well remember when the May horse sale was held outside at the Billings Livestock on First Avenue North and would run for three days," Bill Parker recalled in his BLS Market News column in the April/May 2015 BLS Horse Sale Update. "It was an exciting time for a kid who always knew he was going to wind up in the horse sale business. Tige Thompson and Bob Thomas were the auctioneers and Ralph Cunningham did the commentary. I could listen to them all day."

Bill Parker had a heart for the good "Montana Ranch Broke" kind of horses. He was raised in nearby Lockwood and graduated from Billings Senior High School in 1972. He competed on rodeo teams in college and won a National Intercollegiate Rodeo Association National Championship in team roping. In his first year competing in the Professional Rodeo Cowboy Association (PRCA), Parker earned Rookie of the Year honors in team roping and qualified for his first National Finals Rodeo. He qualified for the NFR once again in team roping and became the first Montanan to ever qualify in tie-down roping. In 1976, Parker earned the title of International Professional Rodeo Association Team Roping World Champion.

Jann Parker was raised a "horse-crazy, barrel racing, 4-H kid" in North Dakota. She sunk spur into Montana's western world in the 1980s at the Northern Rodeo Association (NRA), which still maintains offices inside the PAYS building on Minnesota Avenue. Jann worked as an NRA executive secretary and director for sixteen years. She and Bill met at a Billings Livestock Horse Sale in 1986, and they married on May 5,

Jann Parker, a "horse-crazy, barrel racing, 4-H kid" from North Dakota, met her husband, Bill Parker, at a Billings Livestock Horse Sale in 1986. They married on May 5, 1989, and managed the Billings Livestock Horse Sales from September 1998 until the time of Bill's passing on June 23, 2016, Having taken the reins of the Billings Livestock Horse Sales as the sole manager since 2016—the first woman in the company's history to ever do so—Jann Parker sees her position as nothing out of the ordinary. She carries on, with a void, yes, but also with a responsibility she sees as an opportunity to perpetuate the western way of life. *Photo by Jody L. Lamp, circa 2010.*

1989. Along with competent horsemanship pulsating through both their veins, any observant onlooker might have summarized the Parkers' synchronized skill sets as a symphony—orchestrated down to the last detail with intermittent beauty, boldness, discipline and a heartbeat for the horse industry and its people.

"My business today is my passion, and I'm lucky enough to share that passion with my husband," Jann stated in an interview with author Jody Lamp in March 2010. "We are a team, sharing common goals—to make the horse industry better for both buyers and sellers. There are no time clocks; your gut tells you when it's time to go home."

For the past twenty-plus years, not a sale prevailed without Jann Parker adorning the auction block with in-season floral or fauna arrangement. And during Bill Parker's time at BLS, anyone outside of earshot could hear his voice boom through the block's microphone and sound system, "C'mon! Everybody on the inside! Everybody on the inside! We're going to start the horse sale!" And once inside, the monthly sales commenced with all the gentlemen being asked to "remove cover" for the singing of the National Anthem, which usually featured local legend Noreen Linderman, "The Outlaw Queen," from Red Lodge, Montana, or the Pledge of Allegiance and a speech to buyers and consignors from Bill Parker, which included the terms and conditions of the sale:

> *On the horses that are sold today, Saturday, these horses will be guaranteed to be sound until Monday at noon. Does not guarantee them to stand a pre-purchase examination or veterinary inspection of any kind, doesn't guarantee your best friend's opinion that knows everything about a horse.... But, what it does promise you is that come Monday morning, you'll still be able to trot them around in that gravel parking lot and they won't limp, they'll see out of both eyes, they'll be good in the air meaning they won't heave, and they will not crib. You get one of them home and they don't fit that description.....Absolutely don't call your friend that knows everything about a horse and damn sure don't call your veterinary. Call me! Bring your horse right back here before noon on Monday…and it doesn't take a rocket scientist to tell you if one of them is limping; if you think he's heaving and not breathing right, we'll take him out there in that sand in the bottom end of that arena and lope him around and see if we can make him roar; and if you think he cribs we'll lock him up in one of those wood pens out there underneath that barn and if he sucks air it won't be but five minutes and he'll be locked on to one of the two-by-eights. So, that's the way it works here. I sell about 10,000 horses a year here and my soundness issues are almost non-existent. As buyers, you need to be responsible for the bumps and blemishes on the horses that will not affect their soundness and more importantly, you need to be responsible for the ride-ability of the horses… and shopped the horses so you can make sure you can ride them when you take them home. Other than that, we don't have many rules here at Billings Livestock. We just want each and every one of you to have a good time!*

Having taken the reins of the Billings Livestock Horse Sales as the sole manager since 2016, albeit the first woman in the company's history to ever do so, Jann Parker sees her position as nothing out of the ordinary. She carries on, with a void, yes, but also with a responsibility she sees as an opportunity to perpetuate the western way of life. She continues to make daily conscious efforts to improve the horse sale market and strengthen the industry and future of horses in the United States. "There's always an opportunity to step it up and keep improving, to never settle for just OK and to lead and not follow," Jann wrote in the September 2016 issue of the BLS News & Updates, after both Pat Goggins and her husband, Bill, passed away within six months of each other. "Time will push change on you even when it's not wanted and you're not looking….But so much has remained the same. Our commitment to keep this sale positive, to continue to be leading our industry, and to continue to find ways to just BE BETTER."[227]

And what better example of staying positive and proactive than the year 2020. At the onset of the worldwide pandemic, coronavirus (COVID-19), Billings Livestock Horse Sale manager Jann Parker made the tough decision to postpone the March and April 2020 sales to wait out the storm. It was only the fourth time in twenty-two years of managing the horse sales that required a postponement due to circumstances beyond their control, said Parker. But she reminded their buyers and consignors that Billings Livestock survived other worldwide events and was confident the horse sale would survive COVID-19 too.[228] Intermediately, the company transitioned some of the cataloged consignments to online bidding formats

> *This is big, worldwide, but we will get through it. Changed, certainly, but we will get there. And Billings will be here ready to have a horse sale. Promise.*
> —*Jann Parker, April 1, 2020*

13
FROM COPPER KING TO THE SPORT OF KINGS

MONTANA'S CONNECTION TO THE THOROUGHBRED HORSE RACING INDUSTRY

MARCUS DALY

Had Marcus Daly lived passed his fifty-eighth birthday, one of Montana's "Copper Kings" and his investments in Thoroughbred horses may have escalated western Montana and the Bitterroot Valley to rival and perhaps surpass any Kentucky Bluegrass-region competitors to become America's Thoroughbred Horse Capital. At the time of his death on November 12, 1900, Daly's accomplishments, accolades and escapades were already foundational in Montana's growth and influence in America's development.

Born on December 5, 1841 at Ballyjamesduff in County Cavan, Ireland, Marcus Daly immigrated to New York City and began his young life as a newsboy in 1856. By this time, the sport of horse racing had reached America's eastern shores nearly two hundred years earlier, when King Charles II of England established the first racecourse in his North American possessions.[229] Daly's love and legacy of Thoroughbreds would come later and full-circle to the city where his journey in America started. But first, his restless spirit sent him west toward California, working his way to join other men with a sparkle of gold in their vision.

Known to have a nose for smelling ore veins, Daly established a reputation as a "miner's miner" after arriving at Nevada's Comstock Lode.[230] His skills would attract the attention of a Utah merchandizing concern called the Walker Brothers, which hired and made Daly a general manager to seek mining opportunities north in Montana. While other fortune seekers pursued the Butte Hill's silver deposits at the Alice Mine, thirty-five-year-old Daly was convinced that copper was the reserve that would be the future source of electricity to meet the territory's demand.

Daly's hunch paid off. Breaking off on his own and with new investors associated with and through the William Randolph Hearst family, Daly would further develop the Anaconda Mine to discover some of the richest bodies of copper sulfide the world had ever seen.[231] The "stocky, likeable and gregarious" Irishman financed the construction of a smelter on Warm Springs Creek to process copper ore from the Butte mines.[232] Six years before Montana was admitted into the Union on November 8, 1889, as the forty-first state, Daly filed a town plat near an ample supply of water and wood for his business operations and called the town Anaconda.

Daly's Bitterroot Stock Farm

> *It is the authors' opinion—based on ten years of traveling for research to Virginia, Kentucky, New York, Nebraska, Montana and Wyoming—that Sir Barton's journey to becoming America's First Triple Crown winner began when Marcus Daly purchased the two-year-old horse, Hamburg, from John E. Madden in December 1897.*

Daly became one of the wealthiest men in the country during the time when Butte was the largest city between Minneapolis, Minnesota and Portland, Oregon. His investments back east brought commerce to Montana, specifically to the Bitterroot Valley, in the form a Thoroughbred horse ranch and stables. Founded by Daly, the Bitterroot Stock Farms near Hamilton grew to 22,000 acres of world-class agricultural and horse breeding facilities. Daly believed the thinner mountain air could produce racing stock of great lung capacity and endurance.[233] Along with a stately mansion, he built breeding, racing and training stables with two exercise tracks to accommodate the territory's heat and cold seasons. A 5/8-mile winter track was enclosed and heated, layers with eight inches of loam topped with bark and surrounded with sod.[234]

> *No stables were too good for his horses. No horses were too good for his stables.*[235]

Marcus Daly, born on December 5, 1841, in Ireland, immigrated to America at age fifteen and served as a newsboy in New York City. Twenty years later, Daly's travels and reputation as a "miner's miner" paid off in Montana. The "stocky, likeable and gregarious" Irishman would finance the construction and processing of copper ore mines in Butte, Montana, to become one of the wealthiest men in the country. His investments back East brought commerce to the Bitterroot Valley in the form of a Thoroughbred horse ranch and stables. *Courtesy of the Ravalli County Museum Photo Archive.*

   Daly rivaled the financial means of his Thoroughbred horse counterparts in New York and Kentucky and therefore spared no expense for acquiring a solid bloodline, caring for his steeds, improving his facilities or hiring talented, experienced trainers. Drawing a crowd to Sheepshead Bay Race Course on the eastern portion of Coney Island, Brooklyn, New York, on a scorching Saturday afternoon, September 11, 1897, was a two-year-old robust bay colt named Hamburg.

   Born with a broad blaze and two hind stockings, Hamburg, the son of leading sire Hanover, was foaled in 1895 on the lush Elmendorf Farm near Lexington, Kentucky. He had been purchased as a weanling by horse trainer John E. Madden for $1,200. With Madden's training, Hamburg won twelve races in sixteen starts and was on a campaign to become the two-year-old Horse of the Year.

   "Young man, do you own that colt?" asked Marcus Daly of the forty-year-old John Madden, inquiring about Hamburg, who had just won a decisive victory in the Great Eastern Handicap. Daly's Montana home-bred Scottish Chieftain had just won the 1897 Belmont Stakes a few

Growing Montana's Agriculture

Founded by Marcus Daly, the Bitter Root Stock Farms near Hamilton grew to twenty-two thousand acres of a world-class agricultural and horse breeding facility. Daly believed the thinner mountain air could produce racing stock of great lung capacity and endurance. Daly's Scottish Chieftain won the 1897 Belmont Stakes in New York; and to this day, he remains the only Montana-bred horse to ever have won the race. When news of Daly and Chieftain's victory reached Montana, the *Western News* touted: "The victory was especially gratifying from the fact that the great three-year-old was bred and raised on the Bitter Root stock farm—a genuine product of the Bitter Root ozone and bunch grass." *Courtesy of the Ravalli County Museum Photo Archive.*

months earlier that year and to this day remains the only Montana-bred horse to ever have won the race.

Hearing no answer, Daly pressed Madden further, "What's your price?"

Madden was attending to Hamburg's ankles and not showing Daly's questions any interest—until just the right moment. Living by the adage "Better to sell and repent than keep and regret," Madden finally yelled over his shoulder back at Daly, "$50,000!"

"$50,000!!??" It wasn't exactly the most arbitrary number for the likes of Daly to hear. In fact, Daly's friend and copper mining business investor James B. Haggin was one of the leading Thoroughbred breeders in the country at the time and had purchased Elmendorf Farm, where Hamburg was foaled. But by 1897, *no horse* on record had ever been sold for more than $40,000.

After a few months of negotiating with Madden, Marcus Daly was ready to make the most historic purchase for a horse in Thoroughbred racing history, paying $40,001, just one dollar more to beat the record.

Daly was now the proud owner of Hamburg. Quite a remarkable "American Dream" story, considering that Daly had arrived at the New York Harbor from Ireland just forty years earlier at age fifteen. From a newsboy on the streets of New York City, Daly now owned the most expensive horse in America. The sale of Hamburg would influence the course of Thoroughbred horse racing history forever.

Daly took Hamburg to his famed trainer, William Lakeland, at his Brighton Beach Race Course stables, Coney Island, Brooklyn, New York. At the same, Madden took his record $40,001 straight back to Lexington, Kentucky, to invest in horses and prime Bluegrass land. For $30,001, Madden bought 235 acres east of downtown Lexington at Winchester Pike. It was known as the Overton Place, where nineteenth-century Kentucky statesman Henry Clay was said to have been married in the main residence. Now, as the new owner, Madden changed the name to Hamburg Place to honor the two-year-old horse he sold to Daly.

Hamburg was an American Thoroughbred racehorse foaled at Elmendorf Farm near Lexington, Kentucky, in 1895. Born with a broad blaze and two hind stockings, Hamburg, the son of leading sire Hanover, was purchased as a weanling by John E. Madden for $1,200. With Madden's training, Hamburg won twelve races in sixteen starts and was named the American Horse of the Year for 1898. Madden sold Hamburg to Marcus Daly, who paid $40,001 dollars, just one dollar more than the previous record of $40,000. *Courtesy of the Ravalli County Museum Photo Archive.*

Soon after the death of financier August Belmont, founder of the Belmont Stakes, in 1890 Daly hired one of Belmont's most trusted trainers, Sam Lucas, and began purchasing Thoroughbreds that had won big races, spending upward of more than $1 million. One of his best purchases came with four yearlings in 1891 from Belle Mead Farm near Nashville, Tennessee. Among them was a chestnut colt foaled in 1889 from the sire Iroquois, the first American horse to win England's Epsom Derby. Daly paid $2,500 for the colt he named Tammany after the famous democratic organization, New York's Tammany Hall, primarily composed of Irish immigrants who came to America around the same time as Daly.

Daly's horse racing ventures continually made for anecdotal fodder back in the Butte copper mines. His most famous and favorite Thoroughbred, Tammany, won his first stakes race at 60–1 odds. He remained undefeated and continued his campaign at four years old. In a classic East meets West match race, Tammany met and defeated the famous Lamplighter, owned by bookmaking tycoon Gottfried Walbaum at the Guttenburg, New Jersey racetrack by a convincing four lengths. To honor Tammany's victory, Daly stuck to his promise: "If Tammany beats Lamplighter, I'll build him a castle." And so he did. In Montana, the Anaconda and Butte miners lived vicariously through the famed Tammany, who was awarded the American Horse of the Year honors in 1892. Unfortunately, Tammany contracted lung fever and died soon after.

> *Marcus Daly died in New York's Netherlands Hotel on November 12, 1900, of complication from diabetes and a bad heart at age fifty-eight. At the time of his death, Daly was one of the wealthiest men in the country. Although accounts of the day believed Daly's burial place would be in Montana, his remains are interned in the Daly Mausoleum in Greenwood Cemetery in Brooklyn, New York. The remains of his wife, Margaret Price Daly, who died forty-one years later, also lay in rest there.*
>
> *Ironically, after Daly's death, John E. Madden, who had sold the horse Hamburg to Daly for a record $40,001 in 1897, would buy Hamburg back on January 30, 1901, at the Bitterroot Stock Farm Dispersal Sale held by Fasig-Tipton Company at Madison Square Garden, New York City. Madden, serving as an agent for William C. Whitney, purchased Hamburg for $60,000. After being in Montana at Daly's Bitterroot Stock Farm, Hamburg would be sent to La Belle Stud, Lexington, Kentucky, until Whitney's death in 1904. Once again, Hamburg would be sent back to Madison Square Garden's by the W.C. Whitney estate and sold*

In 1886, Marcus Daly purchased the existing Anthony Chaffin homestead, including the farmhouse and had it completely remodeled by 1889 to be the Daly family summer residence south of Hamilton, Montana. Less than ten years later, the home was to be remodeled again to a Queen Anne–style Victorian home and again to a new Georgian Revival–style home. However, Marcus Daly died in New York City on November 12, 1900, before seeing the remodel complete. The Daly Mansion in New York City was located at 725 Fifth Avenue, where present-day Trump Tower exists. *Courtesy of the Ravalli County Museum Photo Archive.*

> *for $70,000 to Harry Payne Whitney, eldest son of W.C. Whitney and father of Cornelius Vanderbilt Whitney, whose fourth wife was the late Marie Louise "Marylou" Whitney.*

Hamburg Place in Lexington, Kentucky, would become John Madden's training haven for the next thirty years and earn him numerous titles, including the "Wizard of the Turf." After training Hamburg, Madden would produce seven other champions from 1901 to 1903 and be named America's leading Thoroughbred horse trainer. By 1910, Hamburg Place had grown to two thousand acres, and Madden shifted his focus from training to breeding. Beginning in 1918 and for eleven consecutive years, John Madden would become America's leading breeder, producing 182 stakes winners, including 5 winners of the Kentucky Derby, 4 Belmont Stakes winners and 5 members of the Hall of Fame—among them a little chestnut colt that Madden named Sir Barton.

From the pairing of the stallion Star Shoot and the dam Lady Sterling by Hanover, Sir Barton was foaled on Hamburg Place on April 16, 1916. He ran four races as a two-year-old but never won. Madden reportedly sold Sir Barton for $10,000 in 1918 to J.K.L. Ross, a Canadian businessman and naval officer. Sir Barton remained a maiden until winning the 1919 Kentucky Derby, Preakness Stakes and Belmont Stakes.

After retiring from racing, Sir Barton's breeder, John Madden of Hamburg Place, partnered with the Jones brothers of Audley Farm near Berrysville, Virginia, as an investor and a Thoroughbred industry mentor. In 1921, the partnership purchased Sir Barton from owner J.K.L. Ross to become the foundation sire for the farm. Sir Barton stood stud in Virginia from 1921 to 1927, and his offspring achieved modest success. By 1928, B.B. Jones had sold Sir Barton to the U.S. Army Remount Service Headquarters Station at Front Royal Virginia, but Sir Barton remained on the farm through the 1932 breeding season.

From Front Royal, Virginia, Sir Barton was eventually sold to Dr. Joseph Hylton of Douglas, Wyoming, in 1933 and was loaned to the U.S. Remount Service Depot at Fort Robinson near Crawford, Nebraska. Sir Barton lived out his last days on Dr. Hylton's ranch in the foothills of the Laramie Mountains near Douglas. He died of colic on October 30, 1937, at age twenty-one and was buried near his paddock with a simple sandstone marker. In 1968, concerned citizens and the Douglas Wyoming Jaycees exhumed Sir Barton's remains and moved them into the town's Washington Park, where a horse statue and a plaque pay tribute to the famous horse and mark his final resting place.

The Billings Livestock Commission Connection to America's First Triple Crown Winner

Thoroughbred horses are as much a part of the west as they are of the East where they originated and have been raised for track purposes for many years. For the past twenty years the Remount stallions in the Western states have contributed a great deal to the horse raising program of the Western ranchers.
—*W.J. Bill Hagen's weekly column,* Western Livestock Reporter, *September 1943*

Art Langman's second annual cataloged three-day saddle horse sale from May 20 to 22, 1946, drew record crowds paying new high prices for saddle and pleasure horses, according to the front-page headlines of the May

When coauthor Jody Lamp met with Scott Langman, grandson of Arthur H. Langman, in 2010, one of the pieces of Billings Livestock Commission history/memorabilia he brought to their first meeting was the May 1946 Art Langman's Second Annual Catalogue. As Lamp quickly thumbed through the pages to see if she would recognize any of the names, she paused on bottom of page 158 when she read the description of the eight-year-old stallion Camel Barton. The footnotes confirmed that Camel Barton's sire, Sir Barton, was a Kentucky Derby, Preakness Stakes and Belmont Stakes winner. *Courtesy of the Scott Langman (grandson of Arthur H. Langman) family, Billings, Montana.*

29, 1946 *Western Livestock Reporter*.[236] Well-broke, gentle horses, particularly Palominos, were in the greatest demand, with "feverish bidding by eastern buyers." The impending railroad strike failed to dampen the attendance as the crowd of three thousand plus arrived from most every state in the union. Buyers from Massachusetts, New York, New Jersey, Maryland, Virginia, West Virginia, Ohio, Pennsylvania and Michigan were represented. In fact, as the article later stated, "practically every state [was] represented by the close of the sale."[237]

Norman Warsinske, owner of the *Western Livestock Reporter*, served as auctioneer and aided Langman's sale to shatter all previous horse sale records in Billings both for attendance and sale receipts. Opening day began with Lots 1 to 350 sold in the outdoor sales arena in three sessions, beginning at 9:00 a.m., 1:15 p.m. and 7:00 p.m., ending the day at 9:30 p.m. For those serious horse sale buyers that stayed through the third day to see Page 158, Lot 1027, an eight-year-old stallion named Camel Barton, the consigner, Dr.

H.C. Hall of Casper, Wyoming, stated that this was their opportunity "to buy one of the highest bred thoroughbreds ever offered."[238]

When author Jody Lamp met with Scott Langman back in 2010, one of the pieces of Billings Livestock Commission history/memorabilia he brought to their first meeting was the May 1946 Art Langman's Second Annual Catalogue. As Lamp quickly thumbed through the pages to see if she would recognize any of the horses' and consignors' names, she paused on bottom of page 158 when she read the description of the eight-year-old stallion Camel Barton. The footnotes confirmed that Camel Barton's sire, Sir Barton, was a Kentucky Derby, Preakness Stakes and Belmont Stakes winner.

In 1919, when Sir Barton won the Kentucky Derby, Preakness Stakes and Belmont Stakes, the three races were not "officially" recognized in the sport of Thoroughbred horse racing as the Triple Crown races. The term had been used by various sportswriters, but it wasn't until December 1950 that the term and title were formally proclaimed at the annual awards dinner of the Thoroughbred Racing Association in New York. Sir Barton was retroactively awarded as the first Thoroughbred in America to win all three races.[239]

14
NEW COMMODITIES CONTINUE TO BE INTRODUCED AND SUCCEED

Pulse Crops Adapt Well

Grant Zerbe and his son Clayton farm near Lustre, and they know pulse crops and especially chickpeas. They were introduced to lentils, peas and chickpeas and incorporated them into their crop rotation of wheat for the nitrogen the commodities put back into the soil. The Zerbes produce the chickpeas under the name ChickWheat. Grant shared that farming can be challenging but it's important to stay progressive and be willing to try new things. There is a high demand right now for the pulses nutritionally, and he's glad he can work on providing the supply.

Kelley Bean

Growing pinto beans in Montana has been a successful venture for the Kelley Bean Company. Kevin Kelley, who heads the company, shared that they have between five and six thousand acres of pintos being grown for them in the state, most raised in the Bridger area and around Terry. He added that there is a lot of competition for acreage, and in Montana the company has a steady base of growers who supply high-quality beans. Kelley acknowledged that it was a good decision to expand business into Montana, "The growers are great people, down to earth, honest and hardworking."

Growing Montana's Agriculture

Left: The Kelley Bean Company processes and stores pinto beans at its plant in Bridger. Montana's pinto bean crop is highly sought after for its quality. *Photo by Melody Dobson.*

Below: Overlooking the Clark's Fork Valley of the Yellowstone River and the Kelley Bean Plant at Bridger, a field of beans grows in front of the plant. Five to six thousand acres of pinto beans are produced in Montana for the Kelley Bean Company. A majority of the beans grown in the state are raised near Bridger and in the Yellowstone Valley near Terry. *Photo by Melody Dobson.*

A History of Montana Agriculture

Honeybees on the High Plains

Stepping into Drange Apiary in Laurel provided insight into an industry that is a big part of the economy of Montana's agriculture. Owners Andy and Jodie Drange were discussing and making plans for the trip he and staff would make to California to take the bees to pollinate almond trees. When back in Montana, their bees are placed in the Yellowstone Valley and are a major part of the agriculture cycle, as the bees pollinate the alfalfa and clover that make up much of Montana's hay crop.

As a beekeeper, a majority of Andy's time is spent keeping his bees healthy and watching for anything that can harm the bees. It's key to the operation and to beekeeping. Besides helping in the management of the business, Jodie works on the marketing. Their honey is processed, prepped and sold throughout the nation.

At the apiary itself, where the hives and the honey are taken care of, Jodie and Andy's son Spencer joins them in working all facets of the business, from tending to the bees, delivering the hives and processing the honey. The business runs all year-round. In the winter, time is spent repairing and prepping new hives for the next season.

During winter months, Montana honeybees make a trip to the southern warmer states to pollinate crops. Drange Apiary of Laurel sets up their bees in California's almond groves starting in January, later moving the bees to work in the apple orchards in Washington state. The bees and their hives are shipped home to Montana on trucks to be set out late spring through the summer. In between seasons, the Dranges' bees are stored in safe spaces in Idaho that are traditionally used to store potatoes. *Courtesy of Jodie Drange, Drange Apiary.*

Growing Montana's Agriculture

A honeybee at work near Laurel. Honey represents an important part of the agriculture economy each year for Montana and is in high demand from consumers who value the healthy benefits the "liquid gold" naturally offers. Keeping bees healthy is a priority. Bees are needed to pollinate the trees in the state's fruit industry in the western valleys. The hay fields and crops from one side of the state to the other are also in need of being pollinated. *Courtesy of Jodie Drange, Drange Apiary.*

A Special Timber Harvest

The timber industry of Montana creates and sustains jobs in the western part of the state and helps stabilize the economies of the rural mountainous regions. Mike Wilson, owner of Whitewood Transport, knows of another kind of timber harvest that doesn't happen that often in Montana. The Kootenai National Forest in northwestern Montana was chosen in 2017 to provide the national Christmas tree for the White House. Wilson applied for the honor to haul the tree and was chosen.

From the moment they harvested the tree, loaded and prepped it, Mike Wilson was there to assist and make sure the tree arrived safely. The tree would make its way to the nation's capital on the U.S. Capitol Christmas Tree Tour. The tree made twenty stops along the way—in Montana, North Dakota, South Dakota, Missouri and Kentucky—before it arrived in Washington, D.C.

Mike Wilson's company has been involved in shipping all types of agricultural products, but this harvest was one he won't forget. The tree chosen for the fifty-third U.S. Capitol Christmas Tree was a seventy-nine-foot Engleman spruce from northwestern Montana.

Distribution

Receiving supplies from and getting products to the markets were of utmost concern throughout every stage of Montana's agricultural development.

In 2017, Whitewood Transport of Billings was selected to haul the nation's Christmas tree to the White House. Mike Wilson, owner of the trucking company, submitted the application for the honor and was involved with the process from start to finish. The tree was harvested, loaded and prepared for hauling from the Kootenai National Forest in northwest Montana and made twenty public relations stops before arriving in Washington, D.C. *Courtesy of Mike Wilson, Whitewood Transport.*

Growing Montana's Agriculture

In the early years, when producers were just growing enough food to feed their families, before they could begin trading and selling, they were faced with the questions of storage and shipping. Equipment and roads were limited. Construction on the first major roads was completed in 1862. The population was growing and wanted all the food they could produce. Wagon freight trains were limited to the amounts they could transport. The arrival of the Northern Pacific Railroad in 1883 helped agriculture production grow to a new level, as producers had a sure way to access the markets.

Homesteaders had their hands full once they arrived at the nearest town close to their land claims. Purchasing, shipping and hauling their supplies and their goods to be sold was a major focus of their everyday life. When possible, they were able to contract with transfer companies to handle the heavy stuff and the muddy clay roads they would have to deal with all year long. Until the arrival of motorized trucks, draft workhorses were necessary for moving agriculture forward.

A Northern Pacific Railroad Rotary Plow clearing the rail at the top of Lookout Pass close to the Montana and Idaho border in 1939. Keeping the lines of distribution open for shipping was and still is a dedicated full-time job. In this photo, several engines were required to keep the mountain pass open. *Courtesy of the Cruzan family.*

A History of Montana Agriculture

Above: T.W. Adams Transport Company was based in Chinook and delivered the equipment and products ordered by the homesteaders and townspeople. Goods were shipped by train and then hauled to landowners by teams of horses. The transfer company's motto was "Everything Anywhere." Thomas Adams homesteaded with family in 1905 on the North Fork of the Milk River. Adams owned and bred quality workhorses for his business. At one time, Adams had forty-four Percheron and Clydesdale horses to make up the four-, six- or eight-horse teams required to pull loads of coal to the businesses and residents in Chinook, plus the heavy loads to the farms and ranches. Adams also operated the City Livery & Stable in Chinook before it burned down. *Courtesy of Floyd Adams, grandson of Thomas W. Adams.*

Opposite: Ed Corcoran, founder of Corcoran Trucking based in Billings, stands in front of one of the first trucks he owned and used to start the company within the mid-1960s. Corcoran began by hauling ag products in the midwestern and western regions of the United States and Canada. Today, Corcoran Trucking has over one hundred employees operating a fleet in excess of seventy-five or more trucks to serve the needs of agriculture, energy, construction and general commerce nationally. *Courtesy of the Corcoran family.*

Ed Corcoran and his wife, Kaye, arrived in Billings on a chilly November day not remembered for the temperature but for the chilling news that came over the radio: President John F. Kennedy had been shot and killed. The nation stopped and mourned for its leader. Kaye remembers it like yesterday.

Starting a trucking company with one truck and developing it into a fleet that has the ability to handle all different goods and services is quite an accomplishment. Agriculture is dependent on timely shipping and receiving. Companies like Corcoran Trucking are necessary to keep ag products and the commodities market competitive.

Growing Montana's Agriculture

The Cocoran Trucking Company was started by Ed and Kaye Cocoran but has now transitioned into the hands of the next generation under the leadership of son John Corcoran with siblings Ed, Paul, Don and Kristi involved as well.

Managing Growth

There were many tough years while trying to make a living in agriculture, but as the state of things developed, so did the state of affairs. Along with the years of drought there were years of rain and years of plenty. As the herd and harvests grew, so did the need to manage the stocks and the bonds. Agriculture was a growing industry in the state. There was the need for good solid agriculture business and management.

A History of Montana Agriculture

Stockman Bank founder, William "Bill" Nefsy. As a rancher, cattleman and banker, Nefsy was dedicated to serving his community and the agriculture industry. His insightful leadership is a true legacy that continues to impact agri-business throughout the entire state of Montana. *Courtesy of Stockman Bank.*

William "Bill" Nefsy is the founder of Stockman Bank who in 1953 bought the old Miles City Bank. Nefsy, his wife and baby daughter, Virginia, moved south of Miles City after buying a ranch in the early 1940s. He continued to grow his ranching operations through hard work, saving money to purchase more land. An opportunity presented itself to purchase the bank when Bill Nefsy went into Miles City to get an ag loan and soon discovered there were no banks in the area making loans for agriculture.

Nefsy realized that by acquiring the bank, he could help his community and region by providing ag loans and fill a need. Today Stockman Bank is recognized as Montana's largest agriculture bank, and the family is still involved in the day-to-day banking operations. William Nefsy's grandson Bill Coffee currently serves as the organization's chief executive officer.

The original bank purchased in Miles City by Stockman Bank Founder, William Nefsy, in 1953. By acquiring the bank, Nefsy could help the entire community and region by providing ag-loans, something not being done by the other banks in town. Today, Stockman Bank is recognized as Montana's largest agriculture bank, and the family is still involved in the day-to-day banking operations. William Nefsy's grandson Bill Coffee currently serves as the organization's chief executive officer. *Courtesy of Stockman Bank.*

A Growing Montana Gives Back

The agriculture industry continued to grow in Montana and stabilize. Farms and ranches were producing substantial harvests by the 1950s, and new mechanical and technological advancements were helping families establish their homes and their communities. Those in the rural communities were dedicated to helping others and supporting those who were actively serving the needs of their fellow man, especially the youth. Many from the ag community enjoyed helping and seeing Franklin Robbie and the Yellowstone Boys and Girls Ranch make a difference and were willing to support by donating the proceeds from acres they would set aside for the ministry or from a calf or two out of next fall's sale.

Franklin Robbie had a vision to help youth who were struggling and felt he was being called and led to start a special program. In the mid-1950s, through the encouragement of many and with the support of a newly founded board of directors, Robbie started the Yellowstone Boys and Girls Ranch to help troubled youth. The organization was able to purchase a ranch on the west side of Billings.

Although the mission and purpose were not created around a ranching theme, the ranch property did help to complement the programming and the activities. Also, the passion of Franklin Robbie to see lives changed resonated with men especially in the agriculture community. Several board members and donors who desired to see the Yellowstone Boys and Girls Ranch succeed came from the agriculture sector.

One of the members from the first board of directors was Norm Warsinske, owner and publisher of the *Western Livestock Reporter*, which covered seven western states. Norm had an affection for YBGR, and being personally acquainted with ranchers all around the country, he gladly shared the purpose and needs of the ranch with his fellow stockmen.

Part 3

EXPERIENCING MONTANA'S AGRICULTURE:
EVENTS OF SIGNIFICANCE

15

THE COOPERATIVE MODEL

There are events that happened during Montana's agricultural timeline that had the capacity to change the economy and the culture. One significant moment happened with the flip of a switch. The day the lights came on in rural America through the power of electricity changed agriculture forever. The state's farms and ranches championed electricity's arrival as the greatest achievement in helping them succeed in their operations. The workload associated with daily chores was eased and the hours of productivity extended. Jobs that used to take two to three days were now reduced to one.

Electricity brought a healthier and more comfortable home environment. Evenings were amplified with radios, and those in the country were able to receive information and news, which helped relieve the feelings of isolation. The ability to use electric refrigeration meant a better diet and improved methods of food preparation. The industry changed the way food was processed and preserved. Business grew in the rural towns as agriculture's demand for supplies and products increased. Those who experienced electricity for the first time remember it like it happened yesterday. At ninety-three years old, Ed Lenhardt, longtime farmer rancher from west Billings, proclaimed, "1937! That's the year electricity arrived at my parents farm." When Nick Schuman, a farmer from Wheat Basin, went to town to look at buying a tractor, his wife kindly reminded him that if he came back with the new piece of farm equipment, there had better be an electric wash machine on the trailer as well.

Driving down Montana's small-town main streets, it's easy to forget that the electricity to the only stoplight was made possible by the efforts of the local citizens during the 1930s and 1940s who formed a cooperative association to bring power to the town. They accessed funds made available through the National Rural Electrification Act (REA) signed by President Franklin Roosevelt in 1936.[240] These loans gave them monies to build the lines to carry the electricity, construct generation plants and install the equipment. Through the REA, almost all of rural America was electrified through the efforts of "electricity cooperatives" that were developed and still continue to serve their communities providing power, leadership and innovation.[241]

In the spirit of neighbor helping neighbor, local citizens came together cooperatively and dedicated time, resources and talents to modernize their agriculture operations and lives. Cooperative associations were created across the state to ensure their communities were not left out and guaranteed the same access to the technology and information revolutions the larger metropolitan areas had been experiencing for decades. Through this cooperative model, effective change was made. The newly formed associations fostered economic impact and developed new opportunities to build the infrastructure that was needed to gain necessary access for services, education and healthcare to sustain healthy living in rural Montana.

> *I commend you for telling the cooperative story in your book. Cooperatives are a part of the legacy for Montana's economy as well as part of the future of Montana. The cooperative model is based on the following Cooperative Principles: Voluntary and Open Membership; Democratic Member Control; Member Economic Participation; Autonomy and Independence; Education, Training and Information; Cooperation among Cooperatives; and Concern for the Community.*
> —Craig Gates, CEO, Triangle Communications and Hill County Electric Cooperative

BENEFITS OF A COOPERATIVE ASSOCIATION

Farmers and ranchers led the way in the successful movement to create cooperatives. They needed a strong united voice to secure better prices at the markets and transportation for their goods and to make sure both the farm and the ranch had access to fuel and the best-quality inputs of seed, fertilizer and stock at the right time for spring's planting. These associations continue

Experiencing Montana's Agriculture

The Laurel Co-op Association store and fuel station in the 1960s. Cooperatives provided members convenient shopping, discounts on fuel and a place to talk farmin' while having a cup of coffee. Laurel Co-op Association would be part of the three-way merger from cooperatives started in the 1930s to create the Town & Country Supply Association. *Courtesy of Wes Burley, general manager, Town & Country Supply Association.*

to play a leading role in agribusiness today, providing the producer supplies, updated agronomy information and demonstrations of the new technology. Besides being a strong advocate, being a member of a local cooperative has its rewards. In good years, profits are shared with members through dividends, and discounts result through the association's ability to purchase goods at bulk or wholesale price, passing the savings on to the members.

Town & Country Supply Association in Laurel was created by the merger of three major cooperatives based in Laurel, Billings and Hardin. Each original cooperative had roots going back to the 1930s. Town & Country's general manager, Wes Burley, commends the three groups for joining forces and has seen how it has worked to gain better results for the producers and members. The needs of farmers and ranchers had grown so much that the cooperative was challenged to figure out solutions to meet the demand and provide the resources required to operate their farm or ranch. To stay on top of new developments, the association separated the cooperative into four divisions: agronomy, energy, farm and ranch supplies and the convenience stores. This allowed them to work with the experts who are researching every day in their respective fields and keep abreast of the latest technology. Recently, Town & Country built a new fertilizer plant in Lockwood east of Billings and began operations January 2018. The efficiency of the new plant saves time and money, and with its location next to the railroad and Interstate Highways 90 and 94, shipping and receiving are more cost effective and customer delivery easier. The new facility, complete with a rail spur from the railroad, is able to hold more than 22,000 tons of fertilizer with the capability to receive 600 tons of raw product per hour and then blend and distribute at 250 tons per hour.[242]

Working Cooperatively

Montana's cooperative effort and movement started in the 1930s during the rural communities' persistent struggle to stay competitive in the agriculture sector. It brought the implementation of the cooperative model, which can be credited for stabilizing and growing agriculture into the state's number one economy. Simply put, it worked and continues to work today.

Yellowstone Valley Electric Cooperative's general manager, Brandon Wittman, when interviewed, quickly acknowledged the founders of the organization:

> The story of our cooperative mirrors that of agriculture in Montana. YVEC was started by a group of pioneers with incredible foresight. They literally pulled themselves up by their bootstraps and brought power to the rural area when no one else would. Their selfless work has had a profound effect on the communities of the Yellowstone River Valley.

Havre is one hundred miles northeast of Great Falls and is the location of Hill County Electric Cooperative and Triangle Communication's offices. Power didn't come into that area until after World War II, and according to Clarence "Fritz" Keller, "you couldn't get copper wire until the war was over." HCEC was founded in 1945, and Keller knows the cooperative's history, having grown up and farmed north of Havre by Simpson, close to the Canadian border. He remembers when the lines were built, "everyone

Experiencing Montana's Agriculture

Above: The Hill County Electric Cooperative was formed in October 1946. Two months later, construction began. By 1948, four hundred miles of line had been built and energized. New offices for the cooperative were opened in May 1951 at 109 West Second Street in Havre. *Courtesy of Hill County Electric Cooperative and Triangle Communications.*

Left: Clarence "Fritz" Keller, from Simpson, has been a leader for the Hill County Electric Cooperative, serving on the board of directors for forty years. He has also represented Montana's electric cooperatives statewide and nationally on the National Rural Electric Cooperative Association. *Courtesy of Hill County Electric Cooperative and Triangle Communications.*

wanted to help the linemen." In April 1949, HCEC proposed telephone service in conjunction with the electric cooperative and, in 1953, founded the Triangle Telephone Cooperative.[243]

Fritz Keller, an active leader for Hill County Electric Cooperative, who for forty years served on the board of directors, understands the electrical needs of the region, and he has worked hard to make sure electric cooperatives are capable of fulfilling those needs in the future. Keller has also represented the state's electric cooperatives on statewide boards and on the national level at the National Rural Electric Cooperative Association.

Communication is crucial to agriculture. Technology and information change fast, and at the risk of falling behind, rural farmers and ranchers in the northeastern part of the state formed the Northeastern Montana Telephone Cooperative Association in 1950 out of necessity to build the lines and infrastructure for the first major telephone service.[244] The member-driven organization changed its name to the Nemont Telephone Cooperative Inc. in 1957 and is primarily known as Nemont. With headquarter offices in Scobey, Nemont is the only carrier providing the telecommunication needs of the area nationally.

From bringing in the telephones to installing internet and cellular services to providing and maintaining secure wireless broadband connection, Nemont's role in rural Montana is indispensable. Its determined focus comes from strong leadership and forward thinking. Mike Kilgore, Nemont's CEO, stressed, "We keep concern for communities in the forefront by supporting the development and advancement of the largely agricultural communities we serve. Whether it's producing safe affordable food, providing clean water, or producing energy, rural communities are working hard day in and day out to share their vital resources with the world."

Production agriculture is dependent on telecommunications. The equipment being utilized today requires the use of Global Positioning Systems (GPS). GPS-guided equipment needs access to digital satellite services and internet connections. Most equipment is computerized and guides itself from the information gathered from sensors and data being captured in the machinery on-site and in progress. Self-equipped computers calculate and measure accurate inputs and outputs, saving money by implementing the best agronomics for each acre in production.

There are still issues to resolve in rural America's communication development. Mike Kilgore acknowledges that not everyone is able to access the internet, as broadband services do not reach every location, or if it does, the signal is just not strong enough. Much of Nemont's Montana service

Experiencing Montana's Agriculture

On November 20, 1952, members of the Northeastern Montana Telephone Cooperative Association signed their first REA loans to build and install a dependable and accessible telephone service for northeastern Montana. Known today as Nemont, the cooperative is a major community leader, providing the advanced telecommunication needs for the present and future in northeastern Montana and areas in the southcentral portion of the state. *Courtesy of Nemont.*

Bringing the telephone lines to northeastern Montana's rural communities provided the ability to become engaged and stay abreast of current news and information affecting daily agriculture decisions. Farmers and ranchers now had immediate access and could respond quicker to the market prices and trends by making a phone call. *Courtesy of Nemont.*

area is remote and encompasses over ten thousand square miles, equivalent to one person per square mile. Nemont serves the northeastern part of the state, including the counties of Daniels, Sheridan, Roosevelt and Valley, and parts of south-central Montana.

Without secure and dependable internet services, students in rural areas required to do homework online may have to drive to a more urban area to find a good connection. Ag-producers wanting to sell or buy online also find themselves searching for a hot spot to connect with enough strength to transmit and receive plenty of data. Another glaring reality is that it may not be possible to make a 911 emergency call from some places. Kilgore clarifies, "A 911 call could be the most important call a person ever makes in their life." Nemont continues to stay dedicated to getting broadband connections to everyone with a fast reliable signal and dependable service, especially in the advent of distance learning and telemedicine playing an important role in rural America's daily life.

Carl Mattson of Mattson Farms lives north of Chester. He conveys how the lack of a good broadband signal interrupts the daily operation of farming, as they cannot get good connections for the farm's cellphone and internet all the time. "Connection to the internet may work fine in the house or office, but once you leave the doorstep the signal is lost," described Mattson. The ability to continue business while in the field is hampered with no Wi-Fi, the term used to refer to the wireless signal provided by broadband networks able to transmit data. Farm equipment needs Wi-Fi, and the farmer rancher needs the access to order parts, get deliveries or call for help when working in the field or with the herd. Besides operating a farm that has been passed down from one generation and is being passed on to the next, the Mattsons are involved in advocating for agriculture locally, statewide and nationally. Carl has served on and chaired several national boards, and his wife, Janice, was the first woman to chair the US Wheat Associates in 2009. From experience, they know how important accessible communication is. Carl and Janice Mattson's son, Vince, served as the 2020 president of the Montana Grain Growers Association.

The cooperative model is being used effectively to navigate the issues facing agriculture in western Montana's Blackfoot River watershed. Jim Stone, longtime area rancher who took over his parent's ag operation in 1985, serves as the chairman of the Blackfoot Challenge. It's a cooperative effort organized to address the challenges facing the resources and livelihoods of the people living in and around the Blackfoot River. When the pressures from years of mining, logging, livestock grazing and recreational use began to cause issues, something had to be done. Stone learned early from his father how important it is to bring everyone to the table to solve the problems facing you. Jim calls it "neighboring up," how he and the other ranchers started figuring out solutions, and by 1993 the Blackfoot Challenge was formed. In time, the private landowners and public agencies began to realize coordinated efforts would help them pass information and create the foundation to begin meeting face to face. A volunteer board was organized, and a seat was created to represent each vested group in the watershed from the private landowners, local business owners and residents, public agencies and conservation groups. The mission of the Blackfoot Challenge was clearly defined at the beginning, and today the twenty-plus-member board and the staff work together to "conserve and enhance natural resources and the rural way of life in the Blackfoot watershed for present and future generations."[245] At the close of an interview with Jim Stone, the chairman summed up a major principle of the Blackfoot Challenge: "Instead of making the mistakes fifty times, learn from each other."

16

A PERSONAL PERSPECTIVE

THE JAKE FRANK STORY

Authors' Note: There is no better way to learn about the impacts the cooperative model had in the rural setting than to hear the story of one who experienced the early years. For decades, native Montanan Karen Yost has dedicated her time, talent and passion for agriculture to her family, state and nation. Within this chapter, Yost, who has served as the president of the Montana Agri-Women and American Agri-Women, shares her personal account and family history about how her German grandparents discovered their new home in Montana through agriculture. She also shares the firsthand account of her father Jake Frank's involvement and leadership with cooperative associations. It is with sincerest respect and great admiration for Karen Yost and her leadership in agriculture that we invite you to read the history of the Phillip and Katherine Frank family and her father's work in developing cooperatives as told by Karen Yost of Billings, Montana.

I am a direct descendent of the distinct culture of Germans from Russia—people who emigrated from Germany in the 1700s into the rich Volga River plains and Black Sea area during the Russian reign of Tsar Catherine the Great, herself a German. The migration followed the welcome call of free land and an opportunity for new beginnings from the utter poverty in Germany after the devastating Seven Years' War (1756–1763).

My ancestors, the Franks and the Buschs, migrated to the Volga region to the colonies of Kautz and Deitel in the 1760s. The promise of free land evaporated as they arrived in the barren, unsettled Steppes. However, after unimaginable hardships, with more than half of the immigrants dying from

Experiencing Montana's Agriculture

The Phillip and Katherine (née Busch) Frank family emigrated from Kautz, Russia, to the United States in 1902 with their young family of three children. Phillip and Katherine Frank settled in the Yellowstone Valley near Park City and are the grandparents of Karen Yost. *Courtesy of Karen Yost.*

harsh, brutal Russian climate in the beginning years, these hardworking Germans eventually developed stable and prosperous communities.

What's extraordinary about these people is that as Germans, they remained living in the nation of Russians. They spoke their native German language and continued their German cultural traditions for many years with little assimilation into the Russian population. By the end of the nineteenth century, countless years of wars waged by Russian tsars coupled with the Bolshevik Revolution resulted in a life of oppression, excessive taxes, robbery of their possessions and extreme poverty.[246] Once again, they reached the point of futility, and America beckoned their discovery.

My grandparents Phillip and Katherine (née Busch) Frank first emigrated from Kautz, Russia, to the United States in 1902 with their young family of three children. They arrived in Baltimore, Maryland, en route to Saginaw, Michigan, where Phillip Frank signed a contract to work sugar beets. Both Phillip and Katherine worked in the sugar beet fields during the summer until they moved to Lincoln, Nebraska, for the winter, where Phillip worked on the railroad and Katherine did custom washing.[247] Two children were born to them in America before they journeyed back to Russia in 1905 to help out with his ailing parents.

With the family estate not yet settled, but the Bolshevik Party in an uproar, Phillip and Katherine left on a perilous trip out of Russia to the United States in 1913. This time, their trip to America was more dangerous, and the family traveled mostly during the night. Before setting sail to America, the Franks were forced to make their way on foot for more than one hundred miles to Bremen, Germany, where they boarded the ship.

The Franks now had eight children ranging in ages from one to eighteen, and Katherine was seven months pregnant. Phillip's brother planned to join them within a couple of weeks, but both he and a sister of Katherine's were sent to Siberia and executed before they were able to follow. (I remain forever thankful that my grandparents made such a bold and dangerous decision that I might be born a U.S. citizen.)

The Franks arrived in the Baltimore port on March 20, 1913. While the United States provided hope for a better future for my family, their existence was not easy. They took the train to Fromberg, Montana, where they arrived to work under another beet contract. The shack they were living in was drafty and cold, not a comfortable place for the birth of their ninth child—my father, Jake Frank.

The entire family, including the children, worked in the beet fields during the summer months and moved into town for the winter (or "they would have frozen to death," according to the remarks of my uncle). Phillip Frank worked as a cobbler, traveling from house to house making shoes, sometimes staying with the families. Perhaps his visits helped in finding a place to rent in Park City. In 1917, he was able to borrow money from the bank to purchase an eighty-acre farm for $8,000. Then, less than three years later, following a serious horse accident complicated with cancer, Phillip Frank died on June 6, 1920. Katherine was left with eight children still living at home in this new country whose language she did not speak and a farm on which she had to make payments. She signed her name only with an "X."

The Jake Frank Story

My father, Jake, born on September 13, 1915, in Fromberg, was only four and a half years old when his father died. He recalled the time to me:

"When Dad died in 1920, I remember the casket on the living room table for the wake. Friends, neighbors and family came to the house and sat around comforting my mother. Mom cried and cried and held me on her lap for a long time until I finally fell asleep there. Then someone must have put me to bed."

The hardships for Katherine Frank and her family were constant. She dealt with the local bankers and was threatened with foreclosure many times. Her children were the farmers, and the only money she had was from selling the cream she got from her milk cows, which had to pay for her growing family and all the farming expenses. At one time, Grandma Katherine was taken to the county jail and fined five dollars for contributing to the truancy of her children, who she had kept home to harvest the sugar beets. She told me of the Ku Klux Klan that marched down her driveway at night with flaming torches, wearing their hoods, to drive the "dirty Dutchmen" from this community. Her children would hide under their beds while Grandma Katherine would pray.

Grandma was determined to raise her children as best as she could. They had no electricity, and she rose early in the morning, about 4:00 a.m., to start the old wood and coal cookstove. They all had to work. Grandma said they might be poor, but they can still be clean. Later on, as Dad took over the farming, many times he wore a white shirt and a necktie in the field.

Money was tight, so the family learned to do without it. "We weren't poor, we just didn't have any money," my dad always said. "We always had our own food—from the garden and the animals." At one time, a widower from Denver came to visit and offered Grandma $1,000 to marry him and move to Colorado. She declined, saying that no one else wanted her children, but she did. So, she thought she would just stay where she was. In the 1930s, to a widow who was struggling to make ends meet, $1,000 was a lot of money.

But there were two things Grandma would not tolerate, liars and thieves, and she maintained a stern line of discipline and respect from all of her children. My father became my grandma's confidant and interpreter. It was only later when he recalled the conversations with the bankers and the creditors that he fully understood the tough negotiations my grandma had to make.

Before Phillip Frank died in 1920, he purchased a registered gray Percheron stallion for $1,100 at 9 percent interest from the Hobart Horse Importing Co., in Greely, Iowa.[248] The stallion threw good colts that the children then raised, broke as teams and sold. Two exceptional progeny of the Percheron stallion were Star and Bill. This intelligent dapple-gray team could haul a load of hay home on their own from Dad's brother's farm, which was twelve miles away in Laurel. The boys would load Star and Bill's wagon with hay and send them down the road on their own while the brothers finished loading the second hay wagon. When the boys brought the second load home, Star and Bill would be waiting for them in the farmyard with the hay. They had carefully pulled the wagon down the twelve miles of the county road, stopped for cars that were going by and finished their journey into the homeplace to wait for someone to take their lines and finish unloading.

One of the first jobs Dad had was to herd milk cows for people who lived in town. He was about seven years old and got paid one dollar per cow per month and herded ten cows. The old "biddy" cows did not stay together, and it was a much-despised job. Grandma used to drive her horse and buggy out to him to bring him lunch. He said he cried a lot on that job, and sometimes Grandma cried with him.

Grandma shared farm equipment and livestock, including a little short-horn bull, with her second-eldest son, Henry, who lived in Laurel. When Dad was about eleven years old, he needed to move the bull to his brother's

Jake Frank was an accomplished horseman and successful at working, leading and training horses that were used for everything in agriculture. When Highway 10 was being built between Park City and Laurel, Jake was paid six dollars per day for a ten-hour day for his time leading and using his own team, Star and Bill, the family's prize Percheron workhorses. *Courtesy of Karen Yost.*

place, but he didn't own a saddle to ride his pony, Laddie. With a rope looped around the bull's horns, my dad rode bareback trying to drive it to Laurel. Only after a couple of miles down the county road, the bull got on the fight and kept trying to turn back to Park City. Dad tried his best to keep the bull from going home and finally got him stopped by wrapping the rope around a fencepost. He didn't know what to do, so he sat down beside of the road and had a good cry. He found a piece of wire lying in the ditch, so he picked it up and sharpened it on a rock. Dad then gradually pulled the rope tighter and tighter on the bull until he finally had him snubbed as close to the post as he could, then drove that piece of wire through the bull's nose and twisted it to make a ring. He ran his rope through the ring, untied the bull and was then able to drive him to his brother's farm. When he got home, his mom got into her precious coin purse and gave Dad a few coins to go to town and buy a nose ring for the bull.

Soon, Dad was hired on to help with the construction of Highway 10 being built between Park City and Laurel. He used the two dapple-gray horses, Star and Bill, and was paid six dollars per day for a ten-hour day. When that segment of the road was finished, the crew moved west toward Livingston. The foreman came to Grandma and asked if Dad and his team could go with them. The foreman promised to watch over him, keep him safe while they were working in Livingston and raise his wages to seven dollars a day. Even though the money would have been a great help to the family, Grandma told him that unless Dad was at home in his bed each night, he could not go.

JAKE FRANK AND REX

Dad had a sorrel, bald-faced half-Belgian horse with a flaxen mane and tail that he raised from a colt and named Rex. The two became constant companions. Dad spent hours working with Rex but never rode him until he was three years old. He would play with Rex, teach him tricks and spend time with him but firmly believed in giving young horses a chance for their bones to finish growing before using them for work.

Once Dad started riding Rex, it was evident that the horse liked to run and had an uncanny common sense. The dedication of the Billings Public Stockyards featured a horse show, so Dad entered with Rex. Though he had never even ridden Rex in an arena, he decided to enter the Stock

Horse Class. The judge told the twenty-seven contestants their pattern, then added, "The faster you go, the higher I'll mark you." When Dad's turn came to run the pattern, he kept up a fast past through the pattern, then, remembering the judge's encouragement to run fast, he opened Rex up to a full gallop toward the judge. The judge turned and ran, but Rex slid to a stop, they pivoted to the right and the left and waited for the judge to dismiss him. The judge sent him in a line by himself. Dad thought, "Well, I must have done something wrong," but come to find out, he placed first! He entered seven events with Rex that day and won five blue ribbons and two reds.

Rex was big, tough, athletic and flashy with his four white stockings. Dad and Rex carried the American flag with pride in the Grand Entries. The duo made a great team picking up bucking horses when Dad started to rodeo. Everyone recognized Rex when they entered the arena, and years later, many people have told me they still remember that big sorrel horse my dad used to pick up on. One time as Dad was picking a cowboy off a bucking horse, the horse's halter caught on Rex's bridle and proceeded to pull it off. Rex kept on working without it. When they got the cowboy safely on the ground and the bucking horse penned, Dad simply dismounted from Rex, put his bridle back on and continued with their job.

Dad shared one of his fondest events during his rodeo "picking-up" days was the privilege he had to work the famous Matched Bronc Riding events of Bill Linderman, All Around Cowboy; and Casey Tibbs, Saddle Bronc Champion in Thermopolis, Wyoming, and Bridger, Montana, in 1954. Dad recalled the day of the Bridger event that brought a steady line of cars for thirty miles on Highway 212 from Laurel to Bridger. Nearly ten thousand spectators came from all over the state. The event was covered by national media, and his photograph was in *Life* magazine. Casey won that match.

Survival

There was no money at the Frank farm for any extras, so the family all pitched in to help each other with what they had. One time, two of Jake's sisters told their mother they would need white blouses for Glee Club. Grandma said, "There is no money for that. You will have to make do with what you have." When their sister-in-law heard of the dilemma, she cut bedsheets to make the girls their blouses. So subsequently, they went without bedsheets.

Dad didn't understand English well when he started school. On his first day of first grade, his teacher said something in English that he didn't understand, so he asked, "What?" She slapped him and knocked him right out of his little stool. His school life was off to a rocky start.

Dad and his siblings had reason to be anxious when classes ended. After school, if they didn't run fast enough, they were sure to get beat up. So, they would hit the school door running when the bell rang. Their sister Mary would try to help by swinging at their attackers with her lard bucket lunch box. The school superintendent used to walk by or watch from his office window, and Dad was astounded that he never raised a hand to intervene.

One summer, a family friend, who was training to box, came to work on the Frank farm to build up his muscles. Boxing was a frequent pastime, and the local men could make some money challenging the "professionals" if they could win a round or two. Dad and his brothers learned how to box, and once they started to use their "skills" a few times at school, they no longer got bullied. Grandma approved. However, when Dad asked permission to join the boxing team, Grandma refused. She said that it was okay to defend himself, but it was not ok to try to hurt someone.

Dad's desire to become a cowboy strengthened as he continued to see cowboys as men of integrity. One cowboy Dad noticed was the only person in a group brave enough to take hold of the local banker's kids and pull them off of a child they were bullying. Nobody else dared to get involved for fear of jeopardizing their finances. But to my Dad, if that was how a cowboy behaved, then he wanted to be a cowboy, too.

So it was not difficult to comprehend his transition from a beet farmer to a rodeo cowboy. While he had farming in his roots, he had horses in his blood. Dad learned to be sympathetic to the underdog. Growing up on the farm had taught him that life is not always fair, but that integrity and standing up for what you believe was of utmost importance. Attending the "school of hard knocks" brought Dad to discover his own personal philosophy, "You need to leave the world in a better place than when you found it." That philosophy was part of what drew him into the culture of cooperatives— the principle of helping each other and forming a coalition to speak in one unified voice.

That principle of success through what he found in the Montana Farmers Educational and Co-Operative Union is one that my dad would hold tightly, and he soon joined.

Attending Farmers Union School in January 1941 in Great Falls was an enlightening and educational experience for Jake Frank. Here he began to cultivate the leadership skills that helped him work with the cooperative associations using the cooperative model. He advocated for his own agriculture livelihood and those of his neighbors. Serving with the Yellowstone Electric Cooperative Association, he helped bring electricity to his farm and ranch and the rural communities of Yellowstone Valley. *Courtesy Karen Yost.*

The object of the (Farmers) Union was the solution of the crop marketing problem to the benefit of both producer and consumer. The Report of the Secretary of Agriculture shows that the farmer receives but 46% of the price paid by the ultimate consumer. The Union proposes to…discourage the credit and mortgage system, to assist the farmer in buying and selling, to encourage more scientific methods of agriculture, to help the farmer in the classification and marketing of their farm produce, to stop gambling in farm products by the Boards of Trade, Cotton Exchanges and other speculators, to more thoroughly systemize the business of farmer similar to other industrial enterprises.[249]

Organization is the beacon light that guides all mankind to a solution of all problems that confront them. You must, if you hope to be successful, inject more sociability into your union. This can only be done by having your wives, sisters and daughters who are eligible to membership join the union. If you once get your ladies interested in your union its future is assured.[250]

In 1916, the National Farmers Union (NFU) had more than three million members.[251] Individual membership fees to the national organization was only sixteen cents, but the quarterly dues to the chapter/state affiliates proved to be a hardship to many farmers who were already financially strapped during the 1930s and 1940s. Dad traveled from door-to-door to sell family memberships to the Montana Farmers Union, which as I recall from our conversations, were fifty cents per family. Even that minimal amount was a tough sell in those depression days. Later on, he became a state director and served as vice-president and president.

The Union's goals were to unite its members for better prices through cooperative marketing, and it expanded into asking for parity for agriculture, establishing a legislative committee, Farmers Union Grain Terminal Association, Farmers Union Livestock Commission, Farmers Union Central Exchange, Farmers Union Insurance, and retail outlets for its members. The resolutions passed at their annual meetings were for progressive changes in state/national improvements for agriculture.

> *(In the Farmers Union) there is no radicalism in its make-up and it has no desire to array one class against the other. It preaches co-operation by day and night. It's one excuse for existence, and its one great aim is to uplift the whole of farm life to a higher plan in social, educational and financial ways.*[252]

These goals and values rang true to my dad, and it was not long before he was thrust into the leadership of both the Farmers Union and the rural electrification organizations. He attended Farmers Union schools in Great Falls, Montana, designed for its leaders. I was intrigued by an autograph in his 1941 FU scrapbook from the Education chair: "Ever since I met you at last year's school, I've appreciated your splendid character and philosophy. It is my sincere hope that my trust and devotion haven't been misplaced. With Sincerity, M.K. Stoltz."

A fair-skinned German, Dad was slight of build and pushed the tape to reach five feet, seven inches tall. He had an infectious smile and a healthy sense of humor. What he lacked in size, he made up for in competitiveness and determination. (And now he knew how to box!) He was a natural leader.

Lessons from Jake

Our kitchen table was a frequent meeting place for state and national leaders of cooperative influence, both public and private. My mother, it seemed, was always "pressing Dad's suit" and packing his little black metal suitcase for yet another trip where he would speak with unyielding passion regarding vital rural issues. He served on the boards of Yellowstone Valley Electric, the Mid-West Electric Consumers Association, the Production Credit Association and the Montana Farmers Union. He was a field manager of the Farmers Union Livestock Commission in Billings, traveling the area to buy livestock for the auctions in Billings. He served in the 35th Session of the Montana legislature in 1957 and testified countless times in Congress and in public meetings.

The growth of Farmers Union helped bring awareness to the need to electrify rural America. While urban communities enjoyed the benefits of electricity, farmers were still doing chores and milking cows in the dark. Farm work was all done by manual labor, and farm wives were sentenced to wringer washing machines and wood and coal kitchen stoves. In 1935, only 10 percent of farm homes had electricity, compared to 70 percent of urban homeowners. The private utility companies maintained that farmers had little use for electricity.[253]

"We wanted electricity and couldn't get it," Jake reported. Two other neighbors along with Dad approached the Montana Power Company; they were willing to pay the whole cost of the power lines if they could only get connected. The lines already ran past these farms, but they still were not allowed to connect the lines. The Rural Electrification Association Act of 1936 provided funds for both building power lines and for farmers to purchase electrical equipment such as wiring the farmhouse, installing a water system and purchasing home electronics.[254]

Dad got involved in helping build the electric cooperative in his area. He called Texas state representative Sam Rayburn, coauthor of the Rural Electrification Act, and Senator George Norris of Nebraska to find out more about how to create a local cooperative. Dad worked diligently with fellow Montana farmers as they helped clear the right of ways with crosscut saws at ten dollars per day, until finally, through the financial backing of Rural Electrification Act, the Yellowstone Valley Electric Cooperative brought electricity to the Frank farm.

And Dad never forgot the lesson he learned from his father buying that Percheron stallion way back in 1917—that good bloodlines made good horses.

Experiencing Montana's Agriculture

Chores reminds us of daily tasks associated with rural America. When electricity arrived on the farm, not only was the house furnished with lights but the barn and the shop were lit up too. Chores were not quite as difficult with power to assist and with lights to see. *Courtesy of Gene Roncka, Willow Point Gallery, Ashland, Nebraska.*

He always found time to train his precious horses. They were as important to him as life itself. I have never seen anyone who could understand the mind of a horse better than my dad. I was privileged to have him for my 4-H horse leader and have him teach me to train horses. When I was letting my horses get a little spoiled, Dad would "take them for a ride" and straighten them out without me knowing. For years I thought I was a great horse trainer!

Dad built a quarter horse business based on tremendous breeding of brood mares and cow-savvy, athletic sires. He competed on them in rodeo—calf roping and team roping—making quite a splash into the rodeo world, out of the world of a beet farmer and into the Montana Rodeo Cowboys Hall of Fame in 2005.

He and I had this private joke between us. All the years I was growing up, when I mentioned my dad in a conversation, a frequent and awed response was, "You're Jake Frank's daughter?" Then, after I received a small notoriety

Jake Frank's legacy and love for horses would continue on to the next generations of his family. Karen Yost, Jake's daughter, enjoyed spending time with her dad and with the horses whenever she could. This photo was taken around the time Karen was crowned Miss Montana in 1968. *Courtesy of Karen Yost.*

in a pageant, my dad laughed and told me someone asked him, "You're Karen Frank's dad?" We both got a kick out of that!

During his "retirement" (which he never really mastered), he began again driving his beloved teams. His competitive spirit never left him, and entering teamster competitions statewide, he placed or won consistently. He won the Montana state Teamster's Championships several years and was inducted into the Teamsters Hall of Fame in 2008. At the Hall of Fame celebration, at age ninety-three, his competitive spirit welled up again when a colleague, ten years his junior, got out on the dance floor to dance to the band.

"Come on, Esther," he said to my mother, "Let's dance." Mom was just not up to it, so he asked his granddaughter Katie. Over the floor, they glided…kind of. He just couldn't stand being outdone!

My dad continued his fervor for making the world a better place until the end of his life, which came later that year. He pursued life with unparalleled passion and a fierce competitive spirit. He was a man of strong convictions and a champion of the less fortunate. He valued his family and friends above all else. He was the last of a rare breed and a cowboy through and through.

There are so many stories left to tell, so much vital history of our great land shaped by him and so many others, so many sacrifices made by these people of the soil and so much gratitude that we owe them, who brought us the blessings and riches we enjoy today. I am thankful that this book will contain some of these stories.

My grandfather's funeral verses, Revelation 7:15–17, tell of the rewards waiting for them in heaven (and for us):

> *And he said….Never again will they hunger; never again will they thirst. The sun will not beat down on them, nor any scorching heat. For the Lamb at the center of the throne will be their shepherd; he will lead them to springs of living water. And God will wipe away every tear from their eyes.*

17
WEATHER

Records

In Montana, the weather is an event in itself. Every workday on the farm or ranch includes checking the weather forecast. On February 9, 1870, President Ulysses S. Grant made monitoring the weather a priority. The establishment of the National Weather Service (NWS) was authorized after a joint congressional resolution directed the secretary of war to oversee the weather.[255] By 1891, it was transferred to civilian activity under the Department of Agriculture.

In the earliest years of our country, the weather records were monitored by individuals. Benjamin Franklin, Thomas Jefferson and James Madison had daily observations and journals maintained. The U.S. surgeon general recorded weather statistics from 1814, the General Land Grant Office from 1817 and the Smithsonian Institute from 1849 on a countrywide basis until the NWS was established. Montana's earliest official records were taken near Fort Benton by a Smithsonian Institution station from October to December 1862.[256] The record for the greatest temperature change in twenty-four hours is held by the state of Montana. The temperature changed 103 degrees from an icy -54 degrees to 49 degrees above zero in Loma on January 14–15, 1972.[257]

Experiencing Montana's Agriculture

Pheasant Run catches the birds flying for cover before a storm makes its way over the prairies. Eastern Montana is home to premier pheasant hunting. Drought and tough winters can take their toll, but pheasants continue to persevere through the harshest elements of the plains. *Courtesy of Gene Roncka, Willow Point Gallery, Ashland, Nebraska.*

THE LAST WORD

Montana's weather was friend or foe, and learning to adapt to its four seasons of intensity continues to fill the history books with pages of detailed stories. Nothing has affected agriculture more than the weather, and oftentimes it would have the last word and affect entire economies. The Bitterroot Valley's entire apple industry was shut down in 1924 by the reoccurring bad weather three years in a row. The apple producers there and industry could never recover.

The blizzard and winter in 1886–87 caused catastrophic losses to livestock on the open range so severe most livestock owners went broke. Drought beginning in 1917 meant the end of the homesteading boom. People left in droves from the prairies of Montana's central and eastern plains, and by the mid-1920s, sixty thousand had exited the state.[258]

Extremes

There have always been extremes in the weather. Plato Pickens, a Huntley Project homesteader from 1907, recalled visiting with Chief Plenty Coup, chief of the Crow tribe, about the weather. Plenty Coup remarked to Pickens that he had seen a summer of drought when the Yellowstone River was only a series of puddles.[259]

The intense weather the pioneers experienced was usually met with "it could have been worse" evaluations; they knew there was little they could do about the situation but keep going or sell out. Sheer determination provided the momentum to see them through. The farmer rancher was often at the mercy of the weather. There was either too much rain or not enough. Damaging winds, hail and freezing temperatures were all part of the elements they would learn to endure. The good years outweighed the bad ones with harvests that brought enough money to keep going. When necessary, they would find jobs off of the farm from delivering the mail or working on the railroad to cleaning and sewing—anything to sustain them until they could be back in the fields. It was important to have dairy cows and chickens, meaning there would be eggs and milk products to sell to the local general store and creamery, which usually raised enough to pay the bank or buy groceries.

Spring run-off from winter snowmelt, long, wet rainy seasons and sudden storms create disastrous floods like the flood of 1918 on the Yellowstone River. The new bridge over the Yellowstone River in Huntley Project was compromised by the high waters and strong current. *Courtesy of the Huntley Project Museum of Irrigated Agriculture Photo Archives.*

Experiencing Montana's Agriculture

Brothers Fred and William Dissly owned and operated the Fergus County Creamery in Lewistown, becoming involved in 1914. Originally from Ohio, they traveled to Montana to consider growing fruit and nuts in the Flathead Valley before going into the creamery business. Creameries were a necessary part of the rural economy. The Disslys purchased milk, eggs and poultry (live and dressed) from the local farmers and were industrious at finding new ways to process and market the raw products. For example, they deshelled eggs, froze and packaged them in thirty-pound cans to be shipped to bakers and food processors. This created jobs that would provide steady income and could balance the farm's operating budget when needed. The Fergus County Creamery was the first plant to pasteurize fluid milk in Montana. It was also the first to install a "Chicken Picker" in the central part of the state and could remove the feathers of 150 chickens in one hour to help the growing poultry business shipping processed chicken to local restaurants, clubs and markets on the West Coast. *Courtesy of Mary Dissly Cruzan, granddaughter of Fred L. Dissly.*

The old general stores of the new rural communities served as lighthouses during the homesteading years. The businesses helped after the storms hit to get things rolling again. Even though it was tough on the proprietors, they knew they were part of a community that was just getting started and it was important they to do their part. As owners they could usually trade for goods since there was a ready market for farm products. In some cases, people were allowed to work off their bill. Skip King, owner of the Billings, Lockwood and Laurel Ace Hardware stores, pointed out these stores had "everything under one roof" and played a critical part in the survival of the homesteaders, especially when the hardships were out of their control. "The general stores were the first businesses homesteaders went in when they first arrived or to get their bearings straight after a weather event."

The land seekers would make arrangements to have their belongings shipped ahead of time simply labeled, "In Care of the General Store." The

In 1906, the Old Huntley General Store was one of the first businesses to meet those seeking land once the Huntley Irrigation Project was completed and opened for homesteading. Mail was received and distributed. Anything that could be ordered and delivered was sent to the general store. Here people stayed connected, heard the news and discovered ways to get through the unforeseen hardships the elements and weather could create in a matter of minutes. *Courtesy of the Huntley Project Museum of Irrigated Agriculture Photo Archives.*

store was also the nucleus for arranging family meetings for members who arrived on the train months later once the land claim had been located and filed. Transportation had to be arranged to travel to the claim site, which could be a short distance away or twenty-five-plus miles from where they arrived. Between the general store and the livery stable, the newcomers found their way to their land. And if the travelers were fortunate, a makeshift building would welcome them made with the supplies from the general store.

"Keep your Eye on the Horizon"

It isn't just a matter of expression, but making sure to watch the clouds forming in the sky was imperative to know if a storm was brewing. The weather could change in the matter of minutes as is shared in the memoirs

Experiencing Montana's Agriculture

The livery stable was an invaluable and dependable service to the new homesteaders. From here they had a reliable means of transportation to reach their new land and to get goods hauled in or out. If needed, or in case of an emergency, they knew they had access to a horse until they could purchase their own. Thomas Adams owned the first livery stable in Chinook and made his horses with buggies or wagons available for those that needed transportation to and from their new claim while they were first arriving and getting started. Adams also owned a transport company that could move the homesteaders' goods to their claims. *Courtesy Floyd Adams, grandson of Thomas W. Adams.*

of Ole and Sigrid Ohlin, written years later by their son Elmer Ohlin, for the book *Sod 'N Seed 'N Tumbleweed: A History of the Huntley Project*.

> *Spring came early in 1909. The homesteaders were in the fields early and looking forward to a good crop. Ole's forty was all plowed and seeding of oats, corn and wheat was done in April, ahead of the spring rains. Ole and Sigrid had never seen such good crops and were very happy. After a week of hot dry weather, July 27 started dead and sultry and about noon clouds were building in the west, soon gathering into a high dark mass. With distant flashes of lightning and the low rumble of thunder, Ole came out of the field as fast as he could run, stopping to put his horses in the barn*

and opening a door for the cows, then on to the house. The wind came first, followed by rain and hail. The storm lasted about twenty minutes. First the windows gave way on the west and north sides of the house, then the rest of the windows blew out from the force of the wind. Hail, glass and water covered the floors; after the storm the sun came out bright and warm and a mist formed over the fields which were white with hail. Nothing was left of the crops, even the corn was shredded and blown away. Sigrid cried, but Ole stood beside his family and said, "Don't worry, we still have the land." Through the winter of 1909 and 1910 the homesteaders helped each other, sharing whatever they had, such as hay, grain, meat and clothing. Eggs and butter were traded for groceries.[260]

Records and stories from other Huntley Project pioneers reflected how the big hailstorm on July 27, 1909, "harvested the crops in fifteen minutes. Everything in the storm's path was destroyed."[261] Farmers knew that the "Great White Combine" had taken a swathe over their land and left little behind but destruction that would take years to restore.

The Aftermath

Huntley Project and the community of Shepherd were hit hard again with a major hailstorm on August 11, 2018. Residents described it as the worst they had ever seen. The storm chose a large path and had started earlier that day in western Montana, gaining momentum by the time it struck much of south-central Montana. Reports claimed it hailed for twenties minutes straight at some locations. Buildings and cars were demolished and any of the crops left to harvest were wiped out.

Harvey Singh, president of SINGH Contracting, specializing in roofing and siding, acknowledged the storm ranked as one of the worst he had seen. As a contractor, Singh knows companies like his are some of the first responders, arriving to help home and landowners clean up the aftermath. SINGH Contracting serves the states in the high plains region: Montana, North Dakota, South Dakota, Wyoming, Colorado and Nebraska. The company's large service area enables his contractors to see some of the worst hailstorms in the nation. Small towns and rural communities are where Singh's company representatives spend a great deal of time working with families that have been there for generations.

Singh reiterates how small agriculture town citizens are vested members of their communities. They've seen tragedies and know hardships. After a storm, it's his company's second nature to work with each community and offer the necessary resources and services to help restore calm and order out of chaos. Singh states:

> *Sometimes it just takes a helping hand, a listening ear or that reassuring reminder that things are going to be all right and get better. It takes time, energy, resources and dedication. In rural communities we know we are working with the legacies of generations. There are families who have been the first ones to homestead the land or build livelihoods' as they developed their farms, ranches, businesses and towns. I feel we have a responsibility to not only provide great service while repairing damage, but to be mindful of preserving that heritage and the traditions.*

Access to supplies in rural areas can be challenging. Singh's company prepares for emergencies in advance. He watches weather forecasts throughout the six-state region regularly, along with other reports:

> *The technology in our business has allowed us to look at the data and forecasts and see the intensity and out comes from the weather to help us be better providers and meet the needs. We now have past historical data and averages with immediate information available from the different areas. Over time this helps us build our "outlook" and plan more effectively. No one can predict the weather and no one ever wants to have to go through its severe effects. They just want to know that they can "fix it" if it happens. We want to be able to say confidently, we can and meet the needs in a timely fashion.*

CONCLUSION

Advocating

As we put a wrap on our narratives and close the chapters on the stories of so many who have gone before us, we are moved by the lives they lived. They were true agrarians who dedicated themselves to see their farms and ranches succeed. These pioneers advocated for and started the first cooperatives, the experiment stations and demonstration farms and gave agriculture a voice in Montana. It's important that we do the same and continue their legacy.

> *Standing up and telling our story has never been more important. With over 7.5 billion people in the world, it is essential everyone understands where their food really comes from…tenacious, dedicated, hardworking farmers and ranchers.*
>
> —Dawn Schooley, territory sales manager, Alltech

Indefatigable

During the course of researching and writing this book, the word *indefatigable* was used many times to describe early pioneers. They were resolved to do what they could with the resources they earned and received and stretched them and themselves as far as they could go. These first settlers developed a constitution to survive. Their journals reflect records of their own discoveries, experiments and investments of their time, talent and treasure. As we delved deeper into the stories of those we wrote about, their characters were often described as "progressive and genial" in spirit and responsibility. It wasn't unusual to learn they may have ridden two days straight to reach a freight wagon before supplies ran out or all night to make it to a special event.

Cooperative

As groups of farmers and ranchers began to migrate to the territory to stake a mining claim or file on a new parcel of land, the "spirit of ingenuity" was more effective if you could work together with someone. Cooperative efforts emerged early, and neighbor helped neighbor get things done.

> *Sawing ice blocks from the river and hauling a wagonload of blocks home to be packed in layers of sawdust in the icehouse was a cooperative effort with neighbors banding together to accomplish the job.*[262]
>
> —*Plato Pickens*

Harvesting usually meant several farms working together on the threshing machines to get one another's harvest completed. On the ranches, neighbors saddled up for the round-up and got the branding done.

A Spirit of Hope

There wasn't any insurance of any kind and there wasn't a lot of assurance, but there was always hope. And, by some good measure, there was always food to go around. After the hailstorms, the droughts, the floods, the blizzards and grasshoppers, there was tomorrow. There was one guarantee they all shared; there would always be plenty of hard work.

> *Those first years were bitterly disappointing, with nothing but hard work to look forward to. It entailed clearing sagebrush and learning to irrigate by trial and error. In 1909, a devastating hailstorm took most of the crops and the only thing harvested was sugar beets. In spite of these setbacks, they had*

Conclusion

good health and enough to eat. They could see that the land would produce, and so they began to have high hopes for the future.[263]

—*Frank Sindelar*

Montana was often referred to as the "next year" country during the early pioneer days of eastern Montana homesteaders and small-town businessmen who still believed that drought on the high plains was unusual and that surely, next year, it would rain.[264]

Faithfulness

Between the lines and paragraphs of the personal journals, diaries and records was the presence of faith. These pioneers endured loss, hardship, fear and sadness from being far from their homelands, families and friends. But time and time again we found these "faith builders" refusing to give into the pains of loneliness and carrying on to bear one another's burdens. They brought their faith with them, and their testimonies are written in the doorposts of the homes, schools and churches they built. The pioneers left memorial markers for the next generations as reminders of the strong foundations created for them and examples of how to get through the tough times by faithfully hanging in there and sticking it out.

The community of Lustre provides a fitting tribute to the unsung heroes who persevered to find a new life and prepared the way for their families, friends and neighbors to share their faith and realize the hopes of a bright future for generations to come. A monument was built and stands near the front of the Lustre Christian High School in northeastern Montana dated July 4, 1966. The inscription reads, "Dedicated to the Homesteaders who ventured to these Montana prairies, in 1916 and later. And under severe hardship established homes, churches, schools and roads which developed into the Lustre Volt-Communities."

Later in 2016, Lustre installed a sign next to the monument to commemorate the 100th anniversary since the homesteaders first arrived and began filing on their homestead claims and committed to proving up. Three words proclaim their unified prayers of faithfulness and thanksgiving:

Lustre Centennial—"By God's Grace"—1916–2016

NOTES

Epigraph

1. https://www.netstate.com/states/symb/birds/mt_western_meadowlark.htm.

Chapter 1

2. The Discovery Writers, Jean Clary, Patricia B. Hastings, Jeanne O'Neill and Riga Winthrop, *First Roots: The Story of Stevensville, Montana's Oldest Community* (Stevensville, MT: Stoneydale Press, 2005), 15.
3. Ibid., 16.
4. Ibid.
5. Ibid., 20.
6. Andrew F. Rolle, "The Italian Moves Westward," *Montana, the Magazine of Western History*, Winter 1966, 13.
7. Ibid., 14.
8. Discovery Writers, *First Roots*, 22.
9. Newton Carl Abbott, *Making of Montana* (Millings, MT: Gazette Printing Company, 1939), p. 117.
10. Founder's Day 175th Anniversary, 1841–2016, Anniversary Calendar, Volunteers of Historic St. Mary's Mission, 31.
11. Discovery Writers, *First Roots*, 25, 26.
12. Abbott, *Making of Montana*, 118.
13. Paul D. McDermott, Ronal Grim and Philip Mobley, *The Mullan Road, Carving a Passage Through the Frontier Northwest, 1859–62* (Missoula, MT: Mountain Press Publishing, 2015), 1.

14. Abbott, *Making of Montana*, 118.
15. Michael P. Malone, Richard B. Roeder and William Lang, *Montana: A History of Two Centuries* (Seattle: University of Washington Press, 1976), 62.
16. Tom Stout, *Montana, Its Story and Biography*, vol. 1 (Chicago: American Historical Society, 1921), 154.
17. Malone, Roeder and Lang, *Montana*, 63.
18. Ibid., 62.
19. *Commercial West, Weekly Journal Covering Banking, Grain and Western Investments, Milling and Grain* 25, no. 1 (March 1914): 24.
20. Founders Day 175th Anniversary Calendar, 14.

Chapter 2

21. Stout, *Montana*, vol. 1, 184.
22. Ibid.
23. Ibid., 186.
24. Ibid., 189.
25. Clyde A. Milner II and Carol A. O'Conner, *As Big as the West: The Pioneer Life of Granville Stuart* (New York: Oxford University Press, 2009), 74.
26. Ibid.
27. "Bannack, Montana," Western Mining History, https://westernmininghistory.com/towns/montana/bannack.
28. Stout, *Montana*, vol. 1, 220.
29. "Virginia City Preservation Alliance, History," Virginia City, Montana, www.virginiacity.com/history.
30. Milner and O'Conner, *As Big as the West*, 101.
31. "Virginia City Preservation Alliance, History."
32. "The Gold Discovery at Helena, Montana," Gold Rush Nuggets, www.goldrushnuggets.com/godiatlachgu.html.
33. Stout, *Montana*, vol. 1, 422.
34. *Copper Camp: Lusty Story of the Richest Hill on Earth* (Writers Project of Montana by Work Projects Administration, Montana Department of Agriculture, Labor and Industry, 1943), 21.
35. Larry Gill, "From Butcher Boy to Beef King," *Montana, the Magazine of Western History*, Spring 1958, 41.
36. Ibid., 43.
37. Ibid., 46.
38. Ibid.
39. Malone, Roeder and Lang, *Montana*, 146.
40. *Grant–Kohrs Ranch, National Historic Site* (Washington, D.C.: U.S. Government Printing Office, 1979).

41. *Grant–Kohrs Ranch National Historic Site, Montana* (Washington, D.C.: U.S. Government Printing Office, 2015), National Park Service, U.S. Department of the Interior, brochure.
42. Gill, "From Butcher Boy to Beef King," 49.
43. Malone, Roeder and Lang, *Montana*, 146.
44. Grant–Kohrs Ranch, Historic Resource Study, http://www.npshistory.com/publications/grko/hrs/contents.htm, chapter 2, 1.
45. Jim Jenks, *A Guide to Historic Bozeman* (Helena: Montana Historical Society Press, 2007), 10.
46. Douglas C. McChristian, *Fort Laramie; Military Bastion of the High Plains* (Norman: University of Oklahoma Press, 2009), 156.
47. Gail Schontzler, "Gamblers and Dreamers: Wild West Town of Bozeman 150 Years Ago," *Bozeman Daily Chronicle*, www.bozemandailychronicle.com.
48. Montana Genealogy, "First Flour Mill," www.montanagenealogy.com/gallatin/first_flour_mill.htm.
49. Ibid.
50. Jenks, *Guide to Historic Bozeman*, 14.
51. "First Flour Mill," 2.
52. Jenks, *Guide to Historic Bozeman*, 14.
53. Ibid.
54. Gail Schontzler, "Nelson Story—Hero, Scoundrel, Legend," *Bozeman Daily Chronicle*, www.bozemandailychronicle.com.
55. Ibid.
56. Kenneth Ross Toole, *Montana: An Uncommon Land* (Norman: University of Oklahoma Press, 1959), 141.
57. Schontzler, "Nelson Story."
58. Jenks, *Guide to Historic Bozeman*, 83.
59. Ibid., 28.
60. Schontzler, "Nelson Story."

Chapter 3

61. Colonel Edward Wentworth, "History of Montana Sheep Industry," address to MT Wool Growers, 1940.
62. Dick Pace, "Henry Sieben: Pioneer Montana Stockman," *Montana, the Magazine of Western History*, January 1979, 4.
63. Ibid.
64. Ibid.
65. Ibid., 5,6.
66. Wentworth, "History of Montana Sheep Industry."
67. Pace, "Henry Sieben," 11.

68. Wentworth, "History of the Montana Sheep Industry."
69. Pace, "Henry Sieben," 15.
70. Sieben Livestock Company, www.siebenlivestock.com.
71. "About Max Baucus," Max S. Baucus Institute, University of Montana, https://www.umt.edu/law/baucus-institute/about-max/default.php.
72. Jenks, *Guide to Historic Bozeman*, 13.
73. Dorothy J. Zeisler, "The History of Irrigation and the Orchard Industry in the Bitterroot Valley" (master's thesis, University of Montana, 1982), 22.

Chapter 4

74. Ibid., 22.
75. William W. Whitfield, *The Bitterroot Valley Apple Boom, A Brief History* (Hamilton, MT: Ravalli County Museum), 1.
76. Ibid.
77. Ibid.
78. William S. Reese, "Granville Stuart of the DHS Ranch 1879–1887," *Montana, the Magazine of Western History*, July 1981, 16.
79. Ibid.
80. Ibid., 21.
81. Toole, *Montana*, 142.
82. Reese, "Granville Stuart," 23.
83. Ibid., 26.
84. Ibid.

Chapter 5

85. *Progressive Men of the State of Montana* (Chicago, 1902), 960.
86. Stanford Stevens, oral history interview, October 12, 2016.
87. Schontzler, "Gamblers and Dreamers," 3.
88. *Progressive Men of the State of Montana*, 960.
89. Nancy Olson, "Rode with John Bozeman," *Forts and Trails Preview*, April 28, 1963.
90. "Tom Kent Dead; Noted Trail Blazer and Indian Fighter Dies Near Big Timber," *Livingston Enterprise*, August 20, 1910.
91. Stevens, oral history.
92. "Tom Kent Dead."
93. *An Illustrated History of the Yellowstone Valley: Embracing the Counties of Park, Sweet Grass, Carbon, Yellowstone, Rosebud, Custer and Dawson* (State of Montana, 1923), 442.
94. *Scotts Bluff, Fort Laramie Treaty of 1851 (Horse Creek Treaty)*, Scotts Bluff National Monument, Nebraska, National Park Service, U.S. Department of the Interior, brochure.

95. Ibid.
96. Ibid.
97. Burton S. Hill, "The Great Indian Treat Council of 1851," *Nebraska History* 47 (1966): 85–110 (article and summary provided online by Nebraska State Historical Society, https://history.nebraska.gov/sites/history.nebraska.gov/files/doc/publications/NH1966Indian_Treaty_1851.pdf).
98. Ibid., 106.
99. "Crow Reservation Timeline: Crow Tribe 2017," Indian Education Division, Montana Office of Public Instruction, https://opi.mt.gov.
100. Nancy Olson, "The Butcher of Fleshing Rock," Pre-Vue, Billings Weekly Entertainment Guide, *Gallatin County Tribune*, October 21–27, 1973, 4.
101. Elizabeth K. McKomas, "Mary Kent," personal memoir, May 1954, 1.
102. Ibid., 4.
103. "Crow Reservation Timeline," 2.
104. P.J. Smith, brief history timeline; Kind/Kindt/Kint/Kent, family genealogy chart, PDF, August 11, 2017, 1.
105. Olson, "Butcher of Fleshing Rock," 5.
106. "Tom Kent Sells 1200 Head of Cattle," *Billings Post*, Spring 1883.
107. Olson, "Butcher of Fleshing Rock," 5.
108. John Spomer, interview, Big Horn County Museum, Hardin, Montana, August 10, 2018
109. Olson, "Butcher of Fleshing Rock, 5.
110. Spomer, Pages Kent Account Journal, interview, Big Horn County Museum, August 10, 2018.
111. Spomer, interview, August 10, 2018
112. "Dusting off the Old Ones: Tom Kent—1903," *Big Timber Pioneer*, November 20, 1924.
113. Jay Stovall obituary, https://www.michelottisawyers.com/jay-stovall, December 15, 2011.
114. Ibid.

Chapter 6

115. Judith Basin County, Montana, www.centralmontana.com/montana_counties/judith_basin_county.
116. Anna Guttormsen Hought, *Anna: Norse Roots in Homestead Soil* (Seattle, WA: Welcome Press), 67.
117. Passport in Time Interviews on the Lewis and Clark National Forest, Ruth Hardenbrook, September 16, 1997, 1, 2.
118. Wikipedia "Preemption Act of 1841," last modified February 27, 2021, https://en.wikipedia.org/wiki/Preemption_Act_of_1841.

119. Centennial Spotlight: William Skelton, Judith Basin Co., MT, www.files.usgwarchives.net/mt/judithbasin/misc/centennial.txt, 1
120. "Fred Renner, Rangeland Rembrandt," *Montana, the Magazine of Western History* 7, no. 4 (October 1957), 21.
121. Ibid.
122. Centennial Spotlight, 2.
123. Ibid.

Chapter 7

124. Guttormsen Hought, *Anna*, 67.
125. Ibid.
126. Ella Waldon Schloss, *History of Rural School Districts of Dawson County, Montana* (self-published, 1973), 33.
127. Ibid.
128. John Vine, 1919–2001 Southview School Alumni Memory Book, Reunion Committee, 2001.
129. Ibid., 2001.
130. Marvin W. Presser, *Wolf Point: A City of Destiny* (Billings, MT: M Press, 1997), 137.
131. Ibid., 5.
132. Ibid., 34.
133. Ibid., 38.
134. Ibid., 71.
135. Alida Vine, *320 or Bust* (self-published, 1974), 1.
136. Minneapolis Steel & Machinery Company, Twin City Tractors, http://twincitytractors.tripod.com.
137. Regina Christopherson Wanderaas, letter, 1955.
138. Montana Fish, Wildlife and Parks, "What Are Aquatic Invasive Species?," www.mt.gov/fishAndWildlife/species/ais.
139. USDA National Agricultural Library, National Invasive Species Information Center, Species Profiles, Emerald Ash Borer, https://www.invasivespeciesinfo.gov.

Chapter 8

140. Guttormsen Hought, Anna, 68.
141. Harry C. McDean, "M.L. Wilson and the Origins of Federal Farm Policy in the Great Plains, 1909–1914," *Montana, the Magazine of Western History* 34, no. 4 (Autumn 1984): 50.
142. Mabel Lux, "Honyockers of Harlem—Scissorbills of Zurich," *Montana, the Magazine of Western History* 13, no. 4 (September 1963): 5.
143. McDean, "M.L. Wilson," 52.

144. Ibid.
145. Ibid.
146. Derek Strahn, "This Is Montana, Homestead Act Launches a New Era in Montana," http://www.umt.edu/this-is-montana/columns/stories/homestead-act-part-three.php.
147. Ibid.
148. Lux, "Honyockers of Harlem," 5.
149. National Park Service, "Homesteading by the Numbers," https://www.nps.gov/common/uploads/teachers/lessonplans/Homesteading%20by%20the%20Numbers.pdf.

Chapter 9

150. Gressenhall Farm and Workhouse, "History of the Binder," https://gressenhallfw.wordpress.com/2014/08/31/history-of-the-binder, 1.
151. Sam Moore, "Ten Agriculture Inventions that Changed the Face of Farming in America," Farm Collector, www.farmcollector.com, 2.
152. Malone, Roeder and Lang, *Montana*, 315.

Chapter 10

153. Donald R. Bosley, "Horsepower," *Montana, the Magazine of Western History* 27, no. 4 (Autumn 1977): 79.
154. Malone, Roeder and Lang, *Montana*, 315.

Chapter 11

155. Trudie Porter Biggers, Homesteading Huntley Irrigation Project, 2018, 6.
156. Huntley Project History Committee, *Sod 'N Seed 'N Tumbleweed: A History of the Huntley Project* (Yellowstone County, Montana, 1977), 10.
157. Ibid., 14.
158. Special Collections Research Center, "Smith–Lever Act," https://www.lib.ncsu.edu/specialcollections/greenngrowing/essay_smith_lever.html.
159. Malone, Roerder and Lang, *Montana*, 315.
160. Ibid.

Chapter 12

161. "Evidence Proves Montana Had First Horse, Local Rotarians Are Informed," *Billings Gazette*, April 1, 1930, 1

162. Clara Owsley Wilson, Alice M. Cusack, Clara Evans and Terry Townsend, *A Child's Story of Nebraska* (Lincoln, NE: University Publishing Company, 1948), 19.
163. International Museum of the Horse, Kentucky Horse Park, Lexington, KY, http://imh.org/exhibits/current/draft-horse-america.
164. Ibid.
165. Ibid.
166. A.F. Buechler and R.J. Barr, *History of Hall County Nebraska* (Lincoln, NE: Western Publishing and Engraving Company, 1920), 281.
167. *Der Herold*, May 25, 1883 (trans. Alton Kraft).
168. Caroline Langman Converse, *I Remember Papa* (Sunland CA: Cecil L. Anderson, 1949), 71.
169. Ibid., 74.
170. Ibid., 75.
171. Fred Langman Sr. obituary, *Grand Island Independent*, December 27, 1924.
172. Converse, *I Remember Papa*, 1949.
173. *Grand Island Independent*, December 8, 1906.
174. Joe Christie, *Seventy-Five Years in the Saddle* (Grand Island, NE: J-Mar Printing, 1976), 7.
175. Buechler and Barr, *History of Hall County*, 585.
176. Ibid.
177. Thomas R. Buecker, *Thomas R. Fort Robinson and the American Century, 1900–1948* (Norman: University of Oklahoma Press, 2002), 24.
178. Wikipedia, "Fort Keogh," retrieved June 25, 2020, https://en.wikipedia.org/wiki/Fort_Keogh.
179. Jody L. Lamp and Melody Dobson, *The Official Commemorative Guide for the BORN TO REIN Documentary Film*, 15.
180. A.F. Buechler and R.J. Barr, *History of Hall County*, 580.
181. Ibid., 252.
182. The Arthur H. Langman family may have lived in Sidney and Scottsbluff, Nebraska, for a brief period. As the *Scottsbluff Star-Herald* reported on Friday, August 17, 1917: "Arthur Langman of Grand Island, one of the proprietors of the Grand Island Horse & Mule company, and one of the best known and experienced horse buyers of the state, was a Scottsbluff visitor Friday, driving by auto from Sidney here. Mr. Langman has word of praise for Scottsbluff and its great progress."
183. Advertisement, *National Wool Grower*, 1924, 3.
184. Advertisement, *Denver Daily Record Stockman*, 1933 Stock Show Edition, January 3, 1933, 50.
185. C.G. Randell and L.B. Mann, "Livestock Auction Sales in the United States, based on a survey by Phil S. Eckert, Montana State College, Bozeman, Mont.," *Farm Credit Administration Bulletin* no. 35, May 1939, 19.
186. Ibid.

187. Billings Livestock Commission, "About Us," http://billingslivestock.com/Cow_Sales/CS_About.html.
188. *Billings Gazette*, Saturday, August 3, 1935.
189. *Billings Gazette*, Friday, December 13, 1935.
190. Ibid., 2.
191. *Billings Gazette*, Tuesday, December 31, 1935.
192. Ibid.
193. CPI Inflation Calculator, https://www.in2013dollars.com/us/inflation/1935?amount=4000000.
194. *Billings Gazette*, Sunday, March 22, 1936.
195. "Livestock Sales Firm Moves to New Location," *Billings Gazette*, January 4, 1976.
196. Ibid.
197. "Plan Livestock Auction Market," *Billings Gazette*, Sunday, February 11, 1934.
198. *Billings Gazette*, Monday, June 1, 1936.
199. "Plans for Livestock Yards Drafted at Conference Here," *Great Falls Tribune*, February 8, 1936.
200. Ibid.
201. "Old Dobbin, Headed for the Last Roundup, Seems to Be Popular Fellow Back East," *Independent-Record* (Helena, MT), August 17, 1934.
202. *Grand Island Independent*, obituaries, Wednesday, December 9, 2015
203. Advertisement, *Billings Gazette*, Two Big Horse Sales, June 1934.
204. Advertisement, *Billings Gazette*, Horses! Horses! Horses!, June 18, 1934.
205. "Friend Throng Cathedral for Torpey Rites," *Grand Island Independent*, April 25, 1940.
206. Information from John Joseph Torpey, August 25, 2016.
207. "John Torpey Succumbs to Heart Attack," *Grand Island Independent*, April 22, 1940.
208. "Horse Sale Prices Hold Steady at Billings Sale," *Western Livestock Reporter*, February 28, 1942.
209. *Western Livestock Reporter*, June 2, 1942.
210. *Western Livestock Reporter*, April 20, 1943.
211. *Western Livestock Reporter*, October 27, 1942.
212. Lamp and Dobson, *A History of Nebraska Agriculture: A Life Worth Living* (Charleston, SC: The History Press, 2017), 107–8.
213. "Funeral Plans for Ivins Walker," *Times Herald* (Montgomery County, PA), July 29, 1939.
214. "Out of State Buyers Active at Billings Auction Market Sale," *Western Livestock Reporter*, February 9, 1943.
215. W.J. "Bill" Hagen, "News of Horses and Horsemen," *Western Livestock Reporter*, April 27, 1943.
216. "Billings Livestock Commission Company, The Pioneer Market—Montana's Largest," *Western Livestock Reporter*, May 25, 1943.

217. "Lyle Devine Acquires Miles City Sales Yard," *Western Livestock Reporter*, October 12, 1943.
218. Ibid., 6.
219. Verda Lucille (Pat) Wyman obituary, *Billings Gazette*, September 12, 2011.
220. War Room, "Harry Truman: Thanksgiving 1945," https://warroom.armywarcollege.edu/articles/thanksgiving-1945/
221. "Billings Stockyards Break Records for Sales, Volume," *Great Falls Tribune*, December 3, 1948.
222. "State Bank Names New Board Member," *Billings Gazette*, November 11, 1952.
223. Wikipedia, "Conrad Burns," last modified March 12, 2021, https://en.wikipedia.org/wiki/Conrad_Burns.
224. "Billings Live Stock Hits 50-year mark," *Great Falls Tribune*, September 30, 1984.
225. "Stockyard Moving to Lockwood Site," *Billings Gazette*, Thursday, May 22, 1975.
226. Linda Grosskopf, *Pat Goggins: As I Saw It* (Billings, MT: Western Livestock Reporter, 2013), 357.
227. Jann Parker, *BLS News & Updates*, August/September 2016.
228. "Going Virtual-How BLS Horse Sales Stays Out Front Despite COVID-19," *Montana Horses Magazine*, April 1, 2020.

Chapter 13

229. *Celebrating 350 Years of Thoroughbred Horse Racing*, Queens Historical Society, Richard Hourahan, curator.
230. Gregory Daschle, "For the Love of Tammany," *Spur*, January/February 1986, 61.
231. Malone, Roeder and Lang, *Montana*, 202.
232. Ibid.
233. Daschle, "For the Love of Tammany," 61.
234. Ibid.
235. C.B. Glasscock, *The War of the Copper Kings* (Riverbend Publishing, 2002), 142.
236. "Annual Langman Cataloged Horse Sale Draws Record Crowds Paying New Highs," *Western Livestock Reporter*, May 29, 1946.
237. Ibid., 4.
238. *Art Langman's Second Annual Cataloged Horse Sale*, May 20–22, 1946, 158.
239. Marvin Drager, "Triple Crown: American Thoroughbred Horse Racing," *Britannica*, https://www.britannica.com/sports/Triple-Crown-American-thoroughbred-horse-racing.

Chapter 15

240. The Living New Deal, "Rural Electrification Act (1936)," https://livingnewdeal.org/glossary/rural-electrification-act-1936, 1.

241. Ibid., 2.
242. Town & Country Fertilizer Plant, "Fertilizer Plant and Distribution Site Development," www.kljeng.com/featured-projects/town-country-fertilizer-plant.
243. Background Data, Hill County Electric Cooperative—Triangle Telephone Cooperative, 1.
244. History of the Nemont Telephone Cooperative Inc., https://www.nemont.com/about-us.
245. Blackfoot Challenge, https://blackfootchallenge.org/history.

Chapter 16

246. Elsie Johnston, *Laurel's Story, A Montana Heritage* (Laurel, MT: Laurel Historical Research Committee: Artcraft Printers, 1979), 27.
247. Randy Rupp, *The Volga Germans* (self-published, 2019), 2.
248. Ibid., 2.
249. Mildred K. Stoltz, *This Is Yours*, (Minneapolis: Lund Press, 1956), 10.
250. Ibid., 21.
251. Ibid., 29.
252. Ibid., 28.
253. Marquis Childs, *The Farmer Takes a Hand* (Garden City, NY: Doubleday & Co., 1952), 33.
254. Ibid., 66.

Chapter 17

255. Barbara Fifer Rackley, "The Hard Winter 1886–1887," *Montana, the Magazine of Western History* 21, no.1 (Winter 1971): 51.
256. Ibid.
257. *Montana Facts*, Tourism and Relocation Magazine, Bitterroot Valley Chamber of Commerce, 51.
258. Strahn, "This Is Montana," 1.
259. Huntley Project History Committee, *Sod 'N Seed 'N Tumbleweed*, 392.
260. Ibid., 370, 371.
261. Ibid., 422.

Conclusion

262. Ibid., 392.
263. Ibid., 441.
264. Presser, *Wolf Point*, 31.

ABOUT THE AUTHORS

Jody Lamp's career in agriculture journalism began in Nebraska. Born and raised in near Minatare in Scotts Bluff County, Jody earned her bachelor's degree in journalism from the University of Nebraska–Lincoln in December 1993 and has minors in psychology, anthropology and history. She worked as a photographer for the *Daily Nebraskan* and as an agriculture writer for the Department of Ag Communications with the Institute of Agriculture and Natural Resources.

After college, she was hired as the agricultural reporter and photographer for the *Beatrice Daily Sun* and later was recruited by Bader Rutter & Associates, the leading agricultural-based public relations and advertising agency in the United States. She moved to Milwaukee, Wisconsin, and lived there for three years before moving to Montana, where she continued to work for the agency. In 2009, Jody opened Lamp Public Relations & Marketing and continues to maintain the office she started at Billings Livestock Commission. Jody and her husband, Mike, recently moved to Mitchell, Nebraska, in May 2015 and are the proud parents of Mark and Jessie. In 2016, Jody was selected to join the Humanities Nebraska Speaker's Bureau.

Jody began collaborating on local and national projects with Montana native Melody Dobson, and together the team created a strategic planning program called "Your One Powerful Voice," designed to help businesses, organizations and individuals collaborate to create a credible resource of influence. Lamp and Dobson worked as the national strategic plan developers for the U.S. Custom Harvesters Inc. and as the national executive co-coordinators for the

About the Authors

Great American Wheat Harvest documentary film—coordinating the strategic plans, fund development for production and public awareness—before launching the American Doorstop Project on July 1, 2015. The American Doorstop Project is a joint-venture agriculture advocacy collaboration to identify, develop, preserve and promote local, regional and national stories that shaped the development of American agriculture through the medium of books, articles, displays, signage and the fine arts of video production.

Melody Dobson served as a signature event coordinator for the National Lewis and Clark Bicentennial's "Clark on the Yellowstone Signature Event" and the "National Day of Honor" from 2004 to 2007. The events commemorated the 200th anniversary of Captain William Clark signing his name at Pompeys Pillar. Melody served as a member of the Pompeys Pillar Historical Association Board of Directors and was inducted on October 1, 2016, into the "Honor Garden" for her dedication to the development and preservation of the monument. Melody comes from a fourth-generation farm and ranch in northeastern Montana and raised her four children—Sheldon, Christopher, Scott and Elizabeth—with her late husband, Terry D. Dobson, in Billings. She has experience in network radio on-air management, as a personality and as news director, as well as in sales. She received her bachelor's degree in communications arts from Montana State University–Billings, formerly Eastern Montana College.

As independent consultants and national project directors, Lamp and Dobson signed a multiple-book publishing agreement with The History Press to produce a series of agriculture history books, beginning with *A History of Nebraska Agriculture: A Life Worth Living*. The American Doorstop Project has partnered with the Montana History Foundation, a 501(c)3 independent nonprofit corporation (www.mthistory.org), so that you, your business or organization can receive a tax-deductible benefit (to the extent allowed by the law) for contributions toward the American Doorstop Project. For more information visit www.AmericanDoorstopProject.com.

About the Artist

To learn more about the artist, Gene Roncka, featured in *A History of Montana Agriculture: A Life of Discovery*, please visit www.americandoorstopproject.com/artist.